Whitestein Series in Software Agent Technologies and Autonomic Computing

Series Editors:
Marius Walliser
Stefan Brantschen
Monique Calisti
Stefan Schinkinger

This series reports new developments in agent-based software technologies and agent-oriented software engineering methodologies, with particular emphasis on applications in the area of autonomic computing & communications.

The spectrum of the series includes research monographs, high quality notes resulting from research and industrial projects, outstanding Ph.D. theses, and the proceedings of carefully selected conferences. The series is targeted at promoting advanced research and facilitating know-how transfer to industrial use.

About Whitestein Technologies

Whitestein Technologies is a leading innovator in the area of software agent technologies and autonomic computing & communications. Whitestein Technologies' offering includes advanced products, solutions, and services for various applications and industries, as well as a comprehensive middleware for the development and operation of autonomous, self-managing, and self-organizing systems and networks.
Whitestein Technologies' customers and partners include innovative global enterprises, service providers, and system integrators, as well as universities, technology labs, and other research institutions.

www.whitestein.com

Advanced Autonomic Networking and Communication

Monique Calisti
Sven van der Meer
John Strassner
Editors

Birkhäuser
Basel · Boston · Berlin

Editors:

Monique Calisti
Whitestein Technologies AG
Pestalozzi Strasse 24
8032 Zürich
Switzerland
mca@whitestein.com

John Strassner
Motorola, Inc.
1301, E. Algonquin Rd.
Schaumburg, IL 60196
USA
john.strassner@motorola.com

Sven van der Meer
Telecommunications Software &
System Group
Waterford Institute of Technology
Carriganore Campus
Carriganore
Co. Waterford
Ireland
vdmeer@tssg.org

2000 Mathematical Subject Classification: 68T05, 68T10, 68T30, 68T50

Library of Congress Control Number: 2007938080

Bibliographic information published by Die Deutsche Bibliothek
Die Deutsche Bibliothek lists this publication in the Deutsche Nationalbibliografie;
detailed bibliographic data is available in the Internet at <http://dnb.ddb.de>.

ISBN 978-3-7643-8568-2 Birkhäuser Verlag AG, Basel – Boston – Berlin

© 2008 Birkhäuser Verlag, P.O. Box 133, CH-4010 Basel, Switzerland
Part of Springer Science+Business Media
Printed on acid-free paper produced from chlorine-free pulp. TCF ∞
Printed in Germany

ISBN 978-3-7643-8568-2   e-ISBN 978-3-7643-8569-9

9 8 7 6 5 4 3 2 1   www.birkhauser.ch

Contents

Contents

III. Networks

Foreword

This book presents a comprehensive reference of state-of-the-art efforts and early results in the area of autonomic networking and communication. The essence of autonomic networking, and thus autonomic communication, is to enable the autonomic component, device or system to govern the set of services and resources delivered at any given time while protecting context-sensitive business goals. An additional challenge is to provide self-governance in the face of changing user needs, environmental conditions, and business objectives. In other words, an autonomic network understands relevant contextual data and changes to those data, and adapts the services and resources it provides in accordance with business-driven policies that protect user and business interests.

Autonomic computing is often described as self-CHOP (self-configuration, -healing, -optimisation, and -protection). Autonomic networking instead focuses on self-knowledge, which is the foundation to build self-governance. Note that self-CHOP functionality is still provided, but the emphasis of autonomic networking is on the foundation to realise self-CHOP, not in the different self-* technologies and benefits.

Given this foundation, the next challenge is how to apply autonomic networking principles in the network on an application-specific basis. Since networks continue to grow increasingly larger and more complex, they become harder to manage efficiently and reliably. The goal is not to eliminate human personnel, but rather to automate the currently numerous manually-intensive tasks that are so error-prone in today's networks. We advocate a formal systems approach in which autonomic devices, components and systems are able to detect, diagnose and repair faults, as well as adapt their configuration and optimise their performance in the face of changing user needs and environmental conditions. Both of these must be done while protecting and healing themselves in the face of natural problems and malicious attacks. Building adaptive and autonomous control directly into the corresponding network elements enables a shift of focus from the technology used by the network elements to the provisioning of next generation converged services.

This special issue explores different ways in which autonomic principles and techniques can be applied to existing and future networks. In particular, this book is divided into three main parts, each of them represented by three papers discussing a particular aspect from industrial as well as academic perspectives.

The first part focuses on architectures and modelling strategies. It starts with a discussion on current standardisation efforts for defining a technological neutral, architectural framework for autonomic systems and networks. This first paper also defines a set of critical system services that Autonomic Networks require and emphasises (along with the other two papers in this section) that a new framework based on standards must be developed to build a new generation of infrastructures (networks and systems) with inherent autonomic capabilities. The second paper examines how a telecommunication company utilises autonomic principles to manage its infrastructure. In particular, this paper focuses on defining semantic information as the basis for knowledge. It defines the need to focus on legacy equipment and services, not just new "clean slate" devices, and advocates the use of software agent technologies. The final paper in this section describes a European effort to model distribution and behaviour of and for (autonomic) network management. While a P2P paradigm was used, this approach is suitable for many different topologies. Its key contribution is the use of a metamodel dedicated to modelling the needs of network management.

Part two of this book is dedicated to middleware and service infrastructure as facilitators of autonomic communications. This part starts introducing a connectivity management system based on a resilient and adaptive communication middleware. A key feature of this approach is its potential for sustained connectivity in the event of path disruptions. The second paper of this part combines the concept of a knowledge plane with real-time demands of the military sector to regulate resources. This paper defines a variant of the Knowledge Plane that uses situatedness as a new management paradigm for gathering, computing and exchanging knowledge and control over a large network. This is followed by a profound discussion on how the management of service access can benefit from autonomic principles, with special focus on next generation networks. This paper concentrates on enabling adaptive connectivity management of nomadic end hosts across heterogeneous access networks using loosely-coupled distributed management functions and control methods.

Part three focuses on how current networks can be equipped with autonomic functionality and thus be migrated to autonomic networks. We start this part by analysing the difference between traditional network management and autonomic network management and learn how the latter enables cross-layer optimisation. This paper emphasises the use of simple and dependable elements that can self-organise to produce more sophisticated behaviours. The second paper shows how a multi-agent system helps to manage a combined MPLS DiffServ-TE Domain. An architecture is described that defines a novel LSP creation strategy that reduces the number of LSPs and hence, the number of signalling operations in the network. Finally, this part concludes with a very interesting approach that applies game theory to autonomically manage the available spectrum in wireless networks in order to improve spectrum efficiency and maximise network revenue. Two different games (revenue-sharing and price) model the spectrum sharing and spectrum trading behaviours between inter-operator radio access networks, leading to a novel bargaining based dynamic spectrum sharing approach that simplifies reaching agreements.

We would like to thank all people who helped us providing this book for you. First of all, all the authors who submitted papers and who made their current research available for this book. Second, our colleagues from Birkhäuser, who gave us the chance of publishing our view on Autonomic Communications. Last not least a special thanks to Roberto Ghizzioli from Whitestein Technologies, who worked very hard over the last summer and constantly pushed us to our limits.

We hope you enjoy and learn from this book as much as we have!

Monique Calisti, Sven van der Meer, John Strassner
Zürich - Waterford - Schaumburg
November 2007

Whitestein Series in Software Agent Technologies, 1–25

Technology Neutral Principles and Concepts for Autonomic Networking

Sven van der Meer, Joel Fleck, Martin Huddleston, Dave Raymer, John Strassner and Willie Donnelly

Abstract. The 2006 MACE workshop [1, 28] has presented the drivers and challenges of Autonomic Networking [2] and fostered an understanding of emerging principles for this new type of networks [3]. In this paper, we present concepts and principles that define a technological neutral, architectural perspective of Autonomic Networks. The work presented in largely based on work within the Architecture team of the TeleManagement Forum (Technological Neutral Architecture, [4]) and joined research work of industrial and academic research teams (for example Ericsson [5], HP & QinetiQ [6] and Motorola [7]). The goal of this paper is to provide the reader with manageability guidelines and architectural patterns leading to the development of manageable autonomic software and communication systems. We present a component-based, distributed system architecture and an associated set of critical system services that Autonomic Networks require. Since we tackle the problem from a technologically neutral angle, this paper will not prescribe a single new technology, but rather provide a means that allows for federating different technologies, each of which offers particular advantages at business and system levels. In particular, it enables business concepts and principles to drive system design and architectures. This may be further implemented using currently available distributed systems information technologies.

Keywords. Technological Neutral Architecture, Autonomic, Contract.

The 1st IEEE Workshop on Modelling Autonomic Communication Environments was part of MANWEEK 2006 (October 25-26, 2006, Dublin, Ireland); organised by the founder of the Autonomic Communications Forum (Radu Popescu-Zeletin, FhG FOKUS), the ACF Chair (John Strassner, Motorola Labs) and the ACF Academic Co-Chair (Willie Donnelly, WIT).

Most of the work presented in this paper is based and extracted from the TMF TNA as described in [4]. Sven is the current editor of this document. Dave, John and Joel are Distinguished Fellows, Joel leads the Architecture Team in the TMF, Martin leads the Service Providers Leadership Council. All co-authors have actively contributed to the TNA [4].

1. Introduction

The technical basis of communication is shifting from typical insular solutions towards interworking environments. Services influence many parts of our daily life and all places people live and work at. With the convergence of data and telecommunications, the complexity, heterogeneity and size of the networks supporting the industry is rapidly increasing. The proliferation of multiple types of smart devices, each with many ways to connect to different networks, complicates not just end-to-end service delivery, but also billing, provisioning and other aspects of creating and managing the lifecycle of services.

From a network operator's or service provider's perspective, Operation Support Systems (OSSs) are no longer capable of easily managing the complex nature of the infrastructure. Future OSSs must take into account not only vastly increased amount of hardware, but also manage the increasing complexity of applications and services in different contexts running on multiple networks.

Autonomic Networks, as an academic concept as well as a commercial opportunity, are seen as a business tool for competitive success overcoming today's roadblocks of innovation by

- managing the increasing business, system, and operational complexity of these environments, and
- reinforcing the ability of the business to determine the specific network services and resources to offer at any given time [22].

Autonomic Networks address the service providers' needs to increase operational efficiency by an order of magnitude and reduce time to market of competitive services. At the same time, software developers and system integrators will find completely new ways of quickly producing profitable solutions. More importantly, services and resources provided by an Autonomic Network will be able to be easily changed by appropriate business goals and policies.

Our work is aimed at the heart of this challenge. We provide the principles and concepts, in a technological neutral way, that allow for a re-thinking on the part of information and communication service providers on how they run and manage their business. We also provide a new way for software developers to embrace these concepts by defining a new way to specify, design and develop management software. The ultimate goal is to define a framework that provides stakeholders all means to dynamically adapt their services and software to the changing needs of customers.

In this paper, we focus not on implementation, but on a logical (*technological neutral*) framework, with appropriate definitions and specifications, that can be published and discussed in order to provide the concepts and tools for (*technological specific*) implementations and deployments. This *architectural* framework will allow for self-aware and self-healing service *creation*. In their *deployment*, these services will be self-adapting, self-optimising and self-configuring. In their operation, these services are envisioned to be self-protecting, self-managing and self-composing. These features, in turn, enable Autonomic Networks to adapt to (rapidly) changing business needs, technological innovations and environmental conditions with no (or very limited) human intervention. With this goal, our work is supporting the development of an Autonomic Communications Framework, whose mission is to support the development of different (autonomic) elements targeted at essential business needs.

The remainder of this paper is organised as follows: Section 2 provides an insight of what MACE has identified as Autonomic Networking. We provide an overview of the general components (or functional blocks) of an Autonomic Network and we introduce the FOCALE architecture, which defines the closed control loop that we are applying to our work. Section 3 provides the bases to define a technological neutral architectural framework, which effectively is a set of concepts and principles using the same terminology and taxonomy to identify challenges in defining a technological neutral architecture. Section 4 uses all basic concepts of section 3 and defines the Distributed Interface Orientated Architecture by means of a conceptual model. This Conceptual Model defines the areas of concern, shows how governance tasks are realised (or supported), and provides a layered approach combining all these aspects. Section 5 then defines the four basic artefacts: Contract, Component, Service, and Policy plus one specification for an Operation. Finally, section 6 gives an overview of the TMF TNA specifications, including framework services and domains. The paper is summarised in section 7 (where we also discuss current work items) and the acknowledgments followed by the list of used references.

2. The Vision of Autonomic Networking

One result of the MACE Workshop [1] was that the presented novel methodologies, architectures, processes and algorithms have been following a similar vision: Autonomic Networking. In essence, this vision describes the ability of a communications system to self-govern its behaviour within the constraints of the business goals that the system as a whole seeks to achieve. To achieve autonomic networking, information and data modelling captures knowledge related to network capabilities, environmental constraints and business goals/policies. Unlike other approaches, Autonomic Networking combines the knowledge from these models with information from a set of ontologies. This produces an augmented set of data structures that, together with reasoning and learning techniques, can be used to reason about network conditions. Knowledge embedded within system models will be used by policy-based network management systems (together with translation/code generation and policy enforcement processes) to automatically configure network elements in response to changing network conditions, environmental changes and changing business goals.

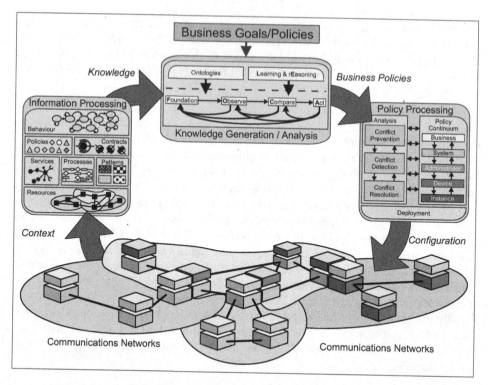

FIGURE 1. Autonomic Network and Closed Autonomic Control Loop [27]

The model-centric approach (Fig. 1) delivers considerable improvements over existing statically configured systems, since it supports the reconfiguration of networks with minimal human intervention. However, to deliver full Autonomic Network capabilities, we believe it is also necessary to introduce processes and algorithms into the system infrastructure to maintain optimal or near-optimal behaviour in terms of global stability, performance, robustness and security (i.e., as developed in [25]).

This yields a high-level approach based on one or more control loops which augments and complements the business models (using for example, eTOM [13] or ITIL® [23]) which are used to govern business tasks. Each activity in the business model can be represented by a set of classes, such as those from DEN-ng [12] that describes the characteristics and features of this activity. This set can then represent the activity lifecycle by being used to construct one or more Finite State Machines (FSM). As the system changes, code is dynamically generated according to the appropriate FSM(s) to protect business goals.

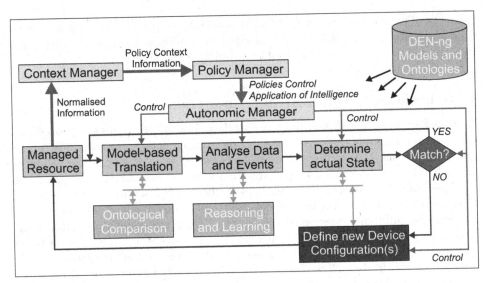

FIGURE 2. FOCALE Architecture – Closed Control Loop (based on [22])

Central to this approach is the presence of one or more system models that abstract the static structure, functionality and dynamic behaviour of the underlying network infrastructure, management functionality and offered services. Also modelled is the governance model of the system, realised as a continuum of policies reflecting business, system, network, device and device instance views [12]. The FOCALE architecture ([3] and [22]) shows how these models are continuously updated in response to the changing operational context of the network, environment, and/or changing business goals. It also describes how this knowledge embodied within the models is utilised to automatically generate configurations that maximise the degree to which the network satisfies business goals given its current operational context.

The basic assumption here is: complexity is everywhere. Thus, FOCALE is first and foremost, a way to manage complexity. Inspired by the autonomic nervous system, FOCALE is using the following analogy: if the autonomic system can perform manual, time-consuming tasks (such as configuration management) on behalf of the network administrator, then that will free the system and the administrator to perform higher-level cognitive functions, such as planning and optimisation. In essence, FOCALE defines the self-adjusting control loop. Inputs to the control loop consist of various status signals from the system or component being controlled, along with policy-driven management rules that orchestrate the behaviour of the system or component. Outputs are commands to the system or components to adjust its operation, along with status towards other autonomic elements. The approach used in FOCALE architecture (cf. Fig. 2) is a policy-driven autonomic control loop incorporating two different loops and allowing for managing legacy components as well as autonomic elements.

3. Technological Neutral Architectural Framework

A technology neutral architectural framework consists of principles and procedures that are used to guide the development of distributed computing solutions. These solutions are based on architectural artefacts, using some or all of these artefacts through the creation and re-use of artefacts retained in a knowledge base. For example, operator deployment teams apply the procedures to identify business needs, model solutions, validate models and build run-time systems. Each team is staffed to fulfil the roles necessary to allow their understanding of the solution to be properly documented. Each team's view of the solution space will be rendered as a set of specifications created by drawing on the artefacts available from the knowledge base. The artefacts retrieved from the knowledge base support building a model (problem, constraints and answer) of the proposed solution. The resulting model is used as the basis for reconciling implementation and realisation decisions once construction on the actual distributed system solution is underway. This methodology, called SANRR for Scope, Analyse, Normalise, Rationalise and Rectify, is described in [14].

The types of artefacts available for use in constructing the solution model are varied, but fall into four general categories:

1. *Process Context* - business process flows, system process plans and process realisation scripts,
2. *Information Context* - business entities, (shared) information models and realisation data models,
3. *Operational Context* - contracts, policies, components, instances and testing systems and
4. *Infrastructure* - technology neutral and technology specific frameworks.

The process and information contexts provide a way to focus on a particular dimension of the solution space. The process context emphasises the high-level behavioural aspects of the solution space while the information context describes specific details regarding the factual aspects (i.e., the static data and dynamic aspects, as well as behaviour and interaction, between components of the solution space. [14]) In FOCALE [22], changes in context cause a potentially new set of policies to be loaded, which adjusts the current governance model to suit the new context. These new policies then control the current functionality that is being offered.

3.1. Terminology

The following is terminology that is used in this document. This terminology is based on the NGOSS TNA specification [4], which incorporates definitions from the NGOSS Meta Model [7], SID resource specifications [9], ODP [10] and TINA [11]:

1. Contract - The central concept of interoperability, providing a specification of operations and behaviour, and exposing functionality contained in a component.
2. Contract Instance - This is a *manageable* runtime manifestation of a *Contract* implementation that provides one or more functions to other runtime entities.
3. Component - A *Component* is defined as a *manageable* software entity that is independently deployable, built conforming to a component software model, and uses *Contracts* to expose its functionality. It represents the unit of deployment in the technology-neutral architecture offering one or more services.
4. Interface - Interfaces represent functionality provided by managed software elements that reside in a *Component*.
5. Service – A group of (one or more) *Contracts* that are managed as a single unit.

3.2. Goal – A Distributed, Interface Oriented Architecture

Autonomic Networking requires an architectural approach addressing software and hardware heterogeneity for both an end-to-end service delivery as well as information systems. A Distributed, Interface Oriented Architecture (DIOA) provides the technology-neutral architectural reference point to satisfy this requirement. A DIOA is defined as follows [4]: a provider entity offers functionality across its interface that involves coordinated behaviour from both the consumer and the provider entities. The set of consumer and provider behaviour, invoked using an interface, is an architectural construct that can be distributed.

The NGOSS technology-neutral architecture, as an example, is specified using a DIOA. In general, a DIOA is already technological neutral. Applications designed using DIOA are not bound directly to any technology. A DIOA should consist of a number of recommendations which form the basis to realise this transparency:

1. An *abstract object* model forms the basis. A concrete object model is derived from it, defining the basic specification elements of an application.
2. A *formal notation* (language) that is used to express the object model, usually combining characteristics for interface definition, managed components and data exchange between components.
3. An *information model* specifying business processes, semantics and behaviour in a generic or domain-specific way.
4. A *Common Communication Vehicle* (CCV) transporting information among components and enabling functionality such as addressing of hierarchies, scoping, filtering and transactions.
5. *Framework services* realising naming, enable mapping of information to e.g., directories and repositories. All services should be accessed in a unified way similar to access of components.
6. *Architectural artefacts* for Business Services: Process, Policy and Security.

In a DIOA, the topology of interfaces is based on the fact that an entity can itself offer one or more interfaces for other entities to bind to and exploit. This hierarchical nesting can go *ad infinitum*. An example of this hierarchy can be found in [4].

With a DIOA providing a technological neutral architectural reference point for design styles, we can achieve optimum business agility and flexibility, services from the entire enterprise need to be re-usable at all layers of the ICT system e.g., customer and resource facing services. Basically, there are three architectural styles a DIOA supports: object-oriented, resource-oriented and service-oriented. A comparison of these styles can be found in [15] and a DIOA related discussion is part of [4].

3.3. Components

Current best practice for DIOAs is to introduce a component model for modelling and implementing run-time entities. DIOA is used for component based software engineering techniques for design, implementation and deployment of solutions constructed [16]. The reminder of this section introduces related work from [16] activities.

Component-based software engineering is concerned with the rapid assembly of systems from components where components and frameworks have certified properties. These certified properties provide the basis for predicting the properties of systems built from components. Components merge two distinct perspectives:

- Component as an architectural abstraction, which express design rules that impose a standard coordination model on all components.
- Component as an implementation, which can be deployed and assembled into larger sub-systems or systems.

A component is an opaque implementation of functionality; a unit of deployment that is subject to third-party composition; and conformant with a component model. The pattern of interaction among different roles, and the reciprocal obligations of components that fill these roles, must be specified. A role is an expected behaviour pattern of an actor in an interaction. Roles are quantified in many ways (constraints, associations and state transition diagrams).

A component model specifies the design rules that must be obeyed by components. These design rules reduce integration problems by removing a variety of sources of interface and architectural mismatch. The rules ensure that system-wide definitions of attributes and behaviour are met, and that components can be easily deployed. These rules cover also how components can be aggregated and/or composed into larger assemblies, and define and fine-tune the external visibility of interfaces.

3.4. Interface, Interface Definition and Service

The ability to integrate components and to develop a market of components depends fundamentally on the notion of contracts. A set of component contracts is a coherent set of functional capabilities (both attributes and operations) of a component that users/clients of those capabilities can rely on.

A DIOA-based system is characterised by the fact that each hardware and/or software entity that provides *services* does so through an interface defined in a way that the corresponding *service* is specified with the following characteristics:

- A description of the *service* (metadata describing its interface and operations, a set of terminations for each operation).
- The behaviour of the *service*: *pre-conditions* define when and how an operation may be invoked. *Post-conditions* define the state that the system is left in for each termination that can be returned when an operation is invoked
- The service must be independently manageable.

Within a DIOA, services are considered to fall into one of two general categories, both of them specified and deployed via components. The differentiation is in the use of the services, not how the services are integrated into the network:

1. Framework Services – providing non-business oriented services to the system; including for example naming/directory, messaging, network time, transaction management/monitoring and policy distribution.
2. Business Aware Services – providing the application level functionality that directly supports the implementation of a business process; including customer record management, customer SLA management, service quality management, billing mediation and rating and discounting.

All services work together in order to deliver the functionality required; framework services have no value add in and of themselves; while business aware services cannot interoperate without the presence of framework services.

3.5. Technology Neutral vs. Technology Specific Architecture

One of the major guidelines of Autonomic Networking is the clear separation between the concepts that are independent of any specific implementation technology (technology-neutral) and the concepts that are specific to one or more particular technologies (technology-specific) [4]. All functionality provided in a TNA is documented in specifications that are neutral with respect to any particular implementation technology. A TNA must be mapped to one or more appropriate Technology Specific Architectures (TSAs). These mappings will leverage industry standard frameworks (e.g., frameworks that support distributed computing, component-based architectures, Service Oriented Architecture) as much as possible.

3.6. Compliance and Certification

Compliance and certification ensure that an application claiming compliance to a DIOA implements and complies with the fundamental principles of it. In order for the NGOSS architecture, as an example, to be ubiquitously deployed, it must be possible to measure the degree of compliance that any given implementation of the NGOSS architecture exhibits. The particular services that are deployed within an

NGOSS based system are not the measure of its compliance. Instead, compliance is measured by verifying if an implementation has implemented the core principles of the NGOSS architecture, as described in [4] and [17]. The following is a listing of the most important of these principles identifying that a compliant DIOA:

1. separates business processes from component implementations,
2. is inherently distributed,
3. uses contracts to communicate and provide functionality,
4. is componentised,
5. is security-enabled,
6. must be policy-enabled,
7. uses shared information and data; h) uses a common repository; and
8. uses a CCV.

4. DIOA Conceptual Model

A DIOA-based system[1] is designed to accomplish one or more purposes, as well as accomplish new purposes through the recombination of existing functionality through contracts, shared data, and a CCV. The system and its purpose can be viewed from different perspectives, whereas each perspective creates its own requirements. However, all perspectives belong to the same basic understanding: A DIOA-based system needs to be prepared to be used and operated in a stable, secure, and efficient way. To guarantee this objective, the system needs to be controlled, administered, and maintained in its entirety, supporting the general aim of the system. This applies to each of its constituent components.

The terms *use* and *operation* describe the part of a system that is seen by users and customers. They are concerned about the system's ability to serve them. A company running a system relies on its efficient *operation* in order to génerate revenue. This operation is supported by controlling the system. The term *control* describes the brief but permanently reoccurring task of keeping the system stable to serve its customers and to generate revenue. This includes, for example, the configuration of system components and the record of data for accounting.

Administration and *maintenance* reflect long term operation and control of a system. This general task is divided into several individual procedures. Administration starts with the permanent monitoring of the system and the logging of all occurring events to analyse the behaviour of its components. The second aim of monitoring is the detection of system failures; a third aim is verification of compliance with contracted behaviour.

4.1. Areas of Concern and Governance Tasks

In Figure 3, four areas of concern are identified for a DIOA. The top block provides the interface to customers enabling them to utilise and manage services.

---

[1]The work described in this section is based on [18], further explored in [5] and detailed as DIOA description in [4].

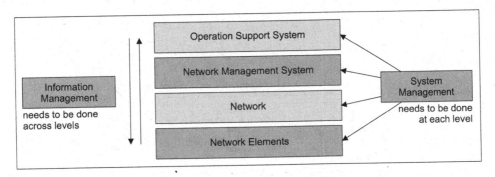

FIGURE 3. Areas of Concern and Supported Governance Tasks

In telecommunications, this is usually provided by an OSS. Services are software assemblies of components that offer functionality and provide access to resources (Network Management System). Resources (Networks and Network Elements) are software and hardware components needed for the provision of services.

Figure 3 shows also the two governance tasks of a DIOA. Information is managed across levels, semantically and syntactically translated upwards and/or downwards; for usage, operation and control. The system is managed at each level through control, administration and maintenance. The assignment of individual activities to one task depends on the system's purpose. This is also true for the separation of the two tasks. The translation of information is supported by management activities and the system's management relies on information translation.

A DIOA also enables the business to be connected to and drive services and resources. The properties of a compliant DIOA previously listed are harnessed by contracts to ensure that the different components of each of the entities shown in Figure 3 interoperate with each other.

Both governance tasks can be described in terms of presentation, specification, submission, re-specification, triggering, queuing, access and execution. This approach is used to model modular client/server applications. It is built upon small and functionally specialised components that can be reused across multiple systems. Each part of this approach provides a specific function in the overall system scope.

The first part focuses on the presentation of information along with the verification of results to support the specification and the submission of individual tasks. The specification answers six questions about a task: *who* (identifier) wants *what* (request) *where* (destination) *when* (schedule) *why* (purpose) and *how* (execution plan). This is, of course, based on the Zachman framework [24]. The answers to *who*, *what*, *when*, and *why* are the basis for a submission that is a complete job specification. The re-specification is responsible for the mapping of *what* towards a set of commands that is needed to be executed to fulfil the purpose. Triggering activates and deactivates jobs based on date and time information, completion of other jobs, or other available data. Queuing provides load balancing and the

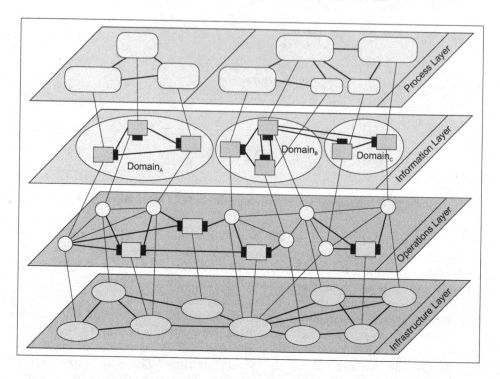

FIGURE 4. DIOA Conceptual Model

prioritising of jobs. Access functions as a mediator between the above layers and the execution layer. It provides interfaces to resources. Execution executes any job that is submitted from submission via access. All described parts are supported by navigation, security, metering and logging.

4.2. A Layered Approach of the Conceptual Model

The layered conceptual model described in this subsection offers concepts and rules for the definition DIOA. The concepts are expressed in form of objectives and requirements. The rules reflect the concept in a multi-layer model. This conceptual model defines rules that need to be followed in order to realise the concepts. Finally, the concepts and the rules can be viewed together to define an architecture.

The main objective of this DIOA layered model is to provide the means of enabling flexibility that is needed to realise Autonomic Networking. It aims for the seamless integration of capabilities in all layers so that management is an embedded capability not independent or tacked on. The aim in Autonomic Networking is to produce extreme flexibility for user applications in the way they may exploit ICT resources in service delivery. The integration serves as a basis to develop service platforms with integrated management facilities that enable all parts of them to be used, controlled, operated, administered and maintained in a unified way.

The second objective is to support the two governance tasks (information management and system management). The model offers mechanisms to map information across identified levels *and* to manage entities within these levels.

We focus on four layers. Each layer is dedicated to a specific problem context. Each problem context describes a dedicated viewpoint to the two areas of concern (information management and system management) and the five terms (use, operation, control, administration and maintenance). The layers are modelled according to the four categories of artefacts introduced earlier in this paper. Furthermore, the layers are used to specify the different types of information that need to be mapped and the different levels of management that is needed.

The conceptual model provides the basis for a specific, yet technological neutral, architecture. The four layers of the conceptual model presented in Figure 4 are:

- Process Layer – governs business tasks. Each business task is then related to a set of classes within the Service Model. A very detailed business process model for telecommunications can be found in eTOM [13] and ITIL® [23].
- Information Layer – combines information models with state machines to create knowledge about the dynamic network environment. For example, the information modelled in SID [20] and DEN-ng [12], combined with state machines, provide a dynamic model allowing for the closed control loop explained in the vision for Autonomic Networking.
- Operations Layer – models resource-facing interfaces and network functionality based on the information models; which will be expressed here in form of (typically vendor-specific) data models. Furthermore, framework and other mandatory services will be provided by this layer.
- Infrastructure Layer – addresses technological specific aspects to specify the mapping from the technological neutral part to the technological specific part.

The shown conceptual model separates business concerns from network technologies. This means, a business application (business processes) is not bound to a particular middleware or management technology. Those technologies become transparent for the business application. The conceptual model allows for substituting technologies without changing the business applications.

5. DIOA Architectural Artefacts

5.1. Contract

A *Contract* is the fundamental unit of interoperability [17]. A contract is a concise specification of functionality - for the purposes of a computation architecture. It is distributed via a component. It is realised by the component execution environment as a computational entity that can be accessed by other computational entities in conformance with the specification as delineated by the contract. Interoperability is important for and within each of the four layers of the conceptual model. For example, *Contract* is used to define a specification of a service to be

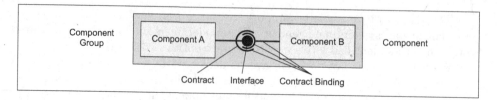

FIGURE 5. DIOA Contracts and Components [17]

delivered, as well as to specify information and code that implement said service. *Contract* is also used to monitor, administer and maintain the service and ensure that any external obligations of the contract (e.g., from an SLA) are met, and to define what measures to take if they are violated in some way.

Contract defines how the service should be used (e.g., invoked) in order to obtain desired results. It is much more than just a software interface specification – it also defines pre- and post-conditions, semantics for using the defined functionality, policies affecting the configuration, use, and operation of the desired functionality and more. In short, *Contract* is a way of reifying a specification of functionality, and guaranteeing the functionality, including obligations to other entities in the managed environment. Thus, it must be viewed as being more than a just container of data or a specification of a set of operations.

Figure 5 shows this concept within a component-enabled DIOA. A component contains one or more contracts. Each of these contracts is realised by an implementation contained and deployed through the contract. The component execution environment instantiates these contained implementations to enable the exposure of a functional interface to the remainder of the component-enabled DIOA. Components may be assembled into (sub) systems (or groups), whereas in this case the (sub) system has internal (not visible to the outside) and external (visible to the outside) contracts. Information exchanged between the functionality contained within components is semantically and syntactically described by a contract. In other words, components are bound by a contract. The binding between contracts can be implicit (realised by an underlying framework) or explicit (realised by special contracts). Policies, not seen in Figure 5, are used to govern the behaviour of a contract.

5.2. Component

A *Component* is an architectural element used to deploy one or more *Contracts*. It represents the unit of deployment in a DIOA [4]. *Components* are containers of contracts (from a minimalist point of view) and containers of contract implementations (from the engineering point of view) for at least two different types:

1. *Contracts* representing non-management functions of the component and
2. *Contract* managing the functions of the component.

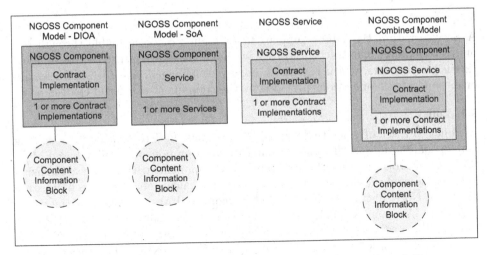

FIGURE 6. DIOA Component – DIOA, SOA and combined Engineering Model [17]

This distinction avoids dictating which contracts the developer of an individual component is required to implement. A component developer may implement those contracts that represent the functionality of their product, be it a self-contained QoS management component (which could be represented by single contract), or an end-to-end customer care system (which could be represented by multiple contracts).

The behaviour the artefacts contained within a *Component* must be manageable. Some *Components* may achieve this manageability by supporting a standard management *Contract* in addition to any supported non-management *Contracts*.

Figure 6 shows a *Component* engineered following the rules of DIOA. The functionality of the *Component* is represented by one or more *Contracts*, thus the *Component* is used to deploy one or more *Contract* implementations. The Component Content Information Block is used to provide information about these contract implementations (most left part of Figure 6).

One way of realising (implementing) a DIOA is using a service-oriented architectural style (SoA), other architectural styles would be object-oriented or resource-oriented. Most of the engineering work within the TM Forum working groups is focused on SoA software development and system engineering. A *Component* implemented following a SoA offers one or more services via its Component Content Information Block (second part from the left in Figure 6). These SoA interfaces must realise a contract according to the general architecture (DIOA) in order to be compliant. Please note that an SoA realisation of DIOA represents a Technological Specific Architecture (TSA). A very good example of an SoA DIOA can be found in [26].

5.3. Service

A Service within the DIOA supplies functionality made available through one or more deployed Contracts. In other words, a *Service* consists of one or more extensible elements, which is created, deployed, managed, and torn down by one or more *Contract* operations provided by the component execution environment.

5.4. Policy

A Policy is a set of rules that are used to manage and to control the changing and/or maintaining of the state of one or more Components. Policy-based Man-. agement controls the state of the system and Components within this system using policies. Control should be realised using a management model such as finite state machines. Note that the emphasis is on "changing and/or maintaining of the state" of a set of Components! This enables behaviour to be choreographed. [12]

Similar to Contracts, Policies are combined to form a continuum [19]. This is necessary in order to relate the needs of different stakeholders to a common problem. The solution provided by this approach, again similar to the concept of Contract, is to define a continuum of policies that enable business, system, implementation, and deployment concepts to be related to each other.

For example: the business view defines the overall goals of an organisation and expresses business policies in a business language. The system view translates this specification into technology and vendor independent terminology. [12]

5.5. Operation

An *Operation* within a Contract is the interaction mechanism by which functionality or parts of functionality is made available. There are two types: *Announcements* and *Interrogations*. All operations are non-blocking. Within a DIOA, the order of delivery is significant, so this order must be guaranteed between different invocations.

An *Announcement* is initiated by the offering functionality, where the offering functionality sends information to the destination functionality, generally without acknowledgement from the destination functionality. Usually, delivery of *Announcements* is not guaranteed (e.g., it is best-effort).

An *Interrogation* comprises two phase of communication; the invocation of this operation from the requesting functionality (*Invocation*) and the response from the offering functionality (*Termination*). Usually, the offering functionality will perform some internal tasks before terminating an Interrogation.

6. NGOSS TNA Specifications

The TMF TNA defines three different types of inter-operating capabilities: *Framework Services* supporting distribution and transparency, *Mandatory Services* supporting decision, sequencing and security, and the general *Business Services* (Figure 7). The CCV is used as a standardised tool to exchange information between

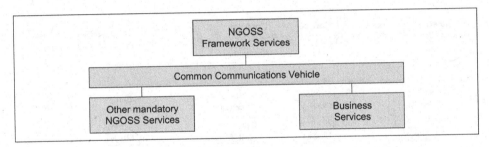

FIGURE 7. NGOSS Technology Neutral Architecture High Level View [4]

these capabilities. The probably best way to think of the CCV is as being a middleware (either remote procedure call based or message based), which organises all functionality that is needed to exchange information (i.e., all basic CORBA [2] services). A Shared Information Model provides information coherence for each component. Underlying this basic view are the essential architectural artefacts.

Figure 8 details Framework Services (operation) and other Mandatory Services (management). Business services are services wrapping legacy applications and services, respectively. The TNA mandates two different types of functionalities:

1. distribution and transparency required for it (Framework Services) and
2. control and coordination of the actions of a system (other mandatory services).

The NGOSS Framework Services provide the infrastructure necessary to support the distributed nature of the NGOSS TNA. In the reminder of this section we provide a brief introduction to these capabilities.

6.1. NGOSS Framework Services

The **Registration Service** provides location transparency. The entities that are considered (from an architectural view) are:

1. the NGOSS System Repository,
2. Shared Information (i.e., Contracts, Processes and Policies),
3. Contract Registrations,
4. Component Content Information Blocks and
5. Contract Instance Registrations.

The Registration Service provides a maintenance interface (for the addition, modification, deletion and browsing of Components, Contracts and Contract Instances) to the NGOSS System Repository.

The **Repository Service** is the heart of distribution transparency. It provides a logical view of all the information about a deployed system. This includes registration information for each business process, Policy, Component, Contract

---

[2]CORBA – Common Object Request Broker Architecture

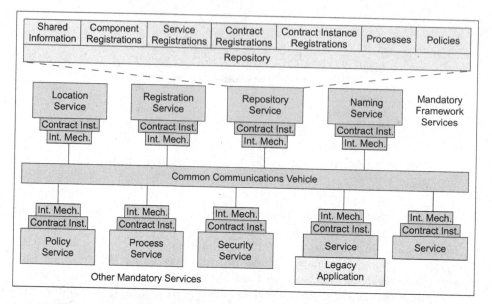

FIGURE 8. NGOSS Technology Neutral Architecture Detailed Views[3] [4]

Implementation and Contract Instance, and the advertising information for each Contract Instance. Although the Repository is often represented as a single instance of a database, this simplification is for illustrative purposes. The Repository may be implemented as a single database, a group of cooperating peer databases, or a hierarchical group of interoperating databases, or a combination of these.

The **Naming Service** is responsible for generating and resolving unique names for the entities contained in the Repository. It interacts with the Registration Services at the time of Component Installation, Contract Registration and Contract Instance Advertisement to assign an identifier that will be used to access, modify, and/or remove the Component, Contract or Contract Instance from the NGOSS system environment, with the Contract Instance Location Services at the time of Contract Instance Location to resolve identified service names into its current location, and with the browsing services supported by the Repository.

Location Services, often built on Naming Services, provide a means to map a request for an object to a particular instance of that object. The NGOSS Location Service is used at run-time to decouple the "hard binding" of consumers and providers to facilitate distribution transparency and the location of available objects.

---

[3] Figure Legend: Contract Inst. = Contract Instance; Int. Mech. = Interface Mechanisms

6.2. Other Mandatory Services

The **Policy Management Service** acts as a supervisor of actions being carried out by all capabilities implemented in an NGOSS system. Hence, this service can apply constraints and/or conditions on what operations are executed, as well as when, by whom and how they are executed. These constraints can be used to impose regulatory, business, enterprise and/or local rules on processing sequences and conditions that are required to be met before (i.e., pre-conditions) and after (i.e., post-conditions and exceptions) any operation. The Policy Management Service is also responsible for the management (activation and deactivation) and monitoring of instantiated Policies.

The **Process Management Service** acts as a conductor or coordinator of activities spanning across the NGOSS Components implementing the Business Services. This Service provides the externalised process control that has been mandated by the NGOSS stakeholders. Ideally, this service executes logic expressed using a means that is different from the implementation language used for the Component. This makes it easier to rearrange and/or alter the business process steps, and then have the Process Management Service rearrange the interaction between the Contract Implementation Instances (if necessary). There may be multiple such Process Management Services within an NGOSS environment, each one used at multiple levels of abstraction in implementing the control structure of a system. Management and monitoring of instantiated processes is an additional responsibility for the Process Management Services deployed within an NGOSS environment.

Security is an essential ingredient in the development of NGOSS systems and should not be considered an item that can be incorporated into a solution at a later date. Rather, Security capabilities must be pervasive throughout the whole NGOSS environment and must be woven into an NGOSS solution from the outset. The **Security Service** provides the infrastructure to support the security requirements.

6.3. Example of NGOSS Communications

Figure 9 shows a typical high-level sequence of interactions that occur between Components in an NGOSS system. The labelled arrows indicate the sequence of interactions that occur from initially registering an instance of a Contract Implementation, through the location of an instance of a desired providing Contract Implementation, resulting finally with the request for functionality and the final response from the provider.

The diagram shows how the decoupled contract instances employ the Framework Services (along with the CCV) to exchange information. The first step in this process is for the Consuming Contract Instance to register with the NGOSS Framework (this is actually done by the consuming Contract Manageability Instance). Next, the consuming Contract Instance queries the Framework Services to locate an instance of the desired Providing Contract Instance (again, via the CCV). Once a Providing Instance is located, a request is formatted and sent. The response comes back in a similar way. It is important to note that there may be

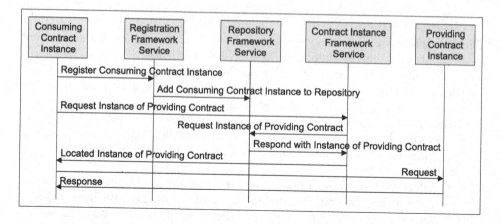

FIGURE 9. Flow of NGOSS Inter-Service Communications [4]

more than one CCV and even more that one CCV technology in an NGOSS system deployment.

Furthermore, the exact nature of the CCV is a technology-specific dependency. That is to say, in some technology-specific implementations, the CCV may be supplied by a specific identifiable runtime entity, whereas in others, it may be an endemic functionality provided transparently by the technology.

This interaction diagram simplifies a number of steps in order to emphasise the decoupling of services and their implementations. Details about the interactions between Contract Instance and the NGOSS Framework Services are found in [21].

6.4. NGOSS Domains

At this point we have defined the artefacts necessary to build the NGOSS domain first depicted at the beginning of this section. Figure 8 shows the build-up and includes a depiction of a legacy application with some (or all) of its capabilities encapsulated with Contract Instance (and Implementations) to allow transparent access by NGOSS client and provider Contract Implementations. The figure also illustrates the NGOSS Repository and shows a sample of the type of information that could be contained in the Repository, including component content, contract, contract implementation, contract instance, policies, process, security and manageability information.

6.4.1. Interoperability between Domains. One of the goals of DIOA and NGOSS is to facilitate the interoperation of multiple domains. Such a requirement could be the result of interoperability agreement between two enterprises, an acquisition, or requirement for interoperability between two different technology domains. To achieve interoperation, the NGOSS TNA defines a concept called *Federation* ([4] and [21]). If the interworking domains are fully NGOSS conformant (i.e., each

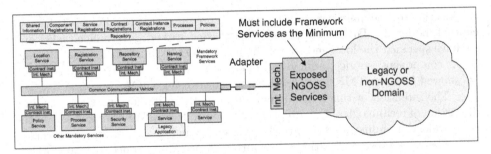

FIGURE 10. Interoperability between two NGOSS compliant Domains[4] [4]

domain implements all of the mandatory services, conforms to implementation model and utilises data models that are semantically equivalent to the SID [20]), interoperation is fairly straight-forward. We only need to specify and implement the federation details (e.g., policies documenting rules for visibility and access) and potentially an *Adaptor* to translate from one CCV technology to another.

6.4.2. Interoperability between an NGOSS Domain and a non-NGOSS Domain. A much more difficult (and common in today's environment) is to interoperate between an NGOSS Domain and a non-NGOSS Legacy Domain. The first step is to identify the functionality that is desired to expose (on each side) and the rules for access and visibility (again for each side). Second, the non-NGOSS Legacy Domain side must deploy Contract Implementations that:

1. wrap the functionality being exposed and
2. implement the Mandatory NGOSS Framework Services.

Finally, the non-NGOSS Legacy Domain must implement either an Interface Mechanism conformant with the CCV, or an *Adaptor* between NGOSS Domain CCV and the distribution technology. This may need to include semantic mapping between data models. Figure 10 provides an illustration of that case.

7. Conclusion

This paper has introduced the vision of Autonomic Networking in form of a short summary of the results of the 2006 MACE workshop [1, 28] based on [2] and [3]. We have discussed a technological neutral architectural framework. With the introduced terminology, we have discussed the idea of a Distributed Interface Oriented Architecture, with the general notion of components, interface, interface definition and service; supported by explanations of two important aspects: compliance and

---

[4]Please note that the left part of this figure is simply a small version of Figure 8. Main aspect of Figure 10 is to show how a non-NGOSS compliant domain can be connected to an NGOSS compliant domain.

certification. In the main part of the paper, we provided a detailed view of the conceptual model of a DIOA (including areas of concern and governance tasks) and we have specified the look and feel of the main DIOA architectural artefacts (contract, component, service, policy and operation). In the final part of this paper, we showed how the TMF TNA has been specified using DIOA concepts.

The intention of this paper is not to focus on an implementation, but on a discussion of technological neutral principles and concepts. We have focused on the communication industry's best practice principles and on standardised concepts.

The work presented in this paper although mature in concept, is incomplete in aspects of its realisation, for example in security solutions. One important aspect is the management of contract instances. A *Contract Instance Management Service* should provide the mechanisms to monitor and log the availability of the contract instances. This service should interact with the repository to update the availability state of each Contract Instance to assure that Contract Instances that are unavailable (as the result of congestion, hardware fault or software fault) are not referred to by the Location Service.

Ongoing work, within the Autonomic Communications Forum[5] and the wider research communities, focuses mainly on the aspect of *Manageability*. We are currently working on use cases for manageability, contract manageability and a reference implementation.

References

[1] W. Donnelly, R. Popescu-Zeletin, J. Strassner, B. Jennings, S. van der Meer (Eds.), *Modelling Autonomic Communications Environments*. Proceedings of 1st IEEE Workshop on Modelling Autonomic Communication Environments (MACE06), Multicon lecture notes, **2**, October, 2006.

[2] N. Agoulmine, S. Balasubramaniam, D. Botvich, J. Strassner, E. Lehtihet, W. Donnelly, *Challenges for Autonomic Network Management*. In Proc. of 1st IEEE Workshop on Modelling Autonomic Communication Environments (MACE06), Dublin, Ireland, October 25-26, 2006.

[3] S. van der Meer, W. Donnelly, J. Strassner, B. Jennings, M. Ó Foghlú, *Emerging Principles of Autonomic Network Management*. In Proc. of 1st IEEE Workshop MACE, Dublin, Ireland, October 25-26, 2006.

[4] TeleManagement Forum (Ed: S. van der Meer), *NGOSS Technological Neutral Architecture*. TMF053, Ed. TNA Release 6.3 (NGOSS R6.1), Document Version 5.7, November, 2006.

[5] R. Carroll, E. Lehtihet, C. Fahy, S. van der Meer, N. Georgalas, D. Cleary, *Applying the P2P paradigm to management of large-scale distributed networks using a*

---

[5]The ACF was founded in 2004 in Berlin with support from the European Commission. In September 2006, the ACF has re-focused towards creating the first international standards body for autonomic communications, by means of unify current thinking in autonomics by creating a new set of ACF sanctioned Autonomic Standards, focusing firstly on the management of systems, and secondly on computing and communications using autonomic mechanisms.

Model Driven Approach. In Proc. of 10th IEEE/IFIP NOMS, Application Session, Vancouver, Canada, April 3-7, 2006.

[6] J. J. Fleck II, S. van der Meer, M. Huddleston, *NGOSS Contract Fingerprints - A Tool for Identifying and Facilitating Re-Use of NGOSS Contracts.* TeleManagement World Americas 2006, Dallas, TX, USA, December 4-7, 2006.

[7] R. Carroll, J. Strassner, G. Cox, S. van der Meer, *Policy and Profile: Enabling Self-knowledge for Autonomic Systems.* In Proc. of 17th IFIP/IEEE Workshop on DSOM, Dublin, Ireland, October 23-25, 2006.

[8] TeleManagement Forum (Ed: J. Strassner), *NGOSS Architecture Technology Neutral Specification – Metamodel.* TMF053D, Ed. TNA Release 6.3, Document Version 1.0, November, 2003.

[9] TeleManagement Forum (Ed: J. Strassner), *Shared Information/Data (SID) Model–Service Overview Business Entity Definitions.* GB922 Addendum 4SO, Ed. NGOSS R6.0, Document Version 6.0, Document Version 2.0, July, 2003.

[10] ITU-T, *Information technology – Open Distributed Processing – Reference model: Overview.* ITU-T X.901 Recommendation, ITU, August, 1997.

[11] TINA-C, *Computational Modeling Concepts.* TINA-C Deliverable, TINA 1.0, Version 3.2, TINA-C, May 17, 1996.

[12] J. Strassner, *Policy-Based Network Management.* Morgan Kaufman Publishers, ISBN 1-55860-859-1, September 2003.

[13] TeleManagement Forum (Ed: M. Kelly), *Enhanced Telecom Operations Map®
(eTOM): The Business Process Framework.* GB921, Ed. NGOSS R6.0, Document Version 6.1, November, 2005.

[14] TeleManagement Forum (Ed: J. Fleck), *The NGOSS Lifecycle and Methodology.* GB927, Ed. NGOSS R6.0, Document Version 4.5, November 2004.

[15] J. Thelin, *A Comparison of Service-Oriented, Resource-Oriented, and Object-Oriented Architecture Styles.* OMG Workshop, Munich, Germany, February 10-13, 2003.

[16] F. Bachmann, L. Bass, C. Buhman, S. Comella-Dorda, F. Long, J. Robert, R. Seacord, K. Wallnau, *Volume II: Technical Concepts of Component-Based Software Engineering.* 2nd Ed., Carnegie Mellon University, Pittsburgh, May 2000.

[17] TeleManagement Forum (Ed: S. van der Meer), *NGOSS Technological Neutral Specification - Contract Description: Business and System Views.* TMF053B, Ed. TNA Release 6.3, Document Version 5.1, November, 2006.

[18] S. van der Meer, *Middleware and Application Management Architecture.* PhD Thesis, Technical University Berlin, October 25, 2002.

[19] S. van der Meer, A. Davy, S. Davy, R. Carroll, B. Jennings, J. Strassner, *Autonomic Networking: Prototype Implementation of the Policy Continuum.* In Proc. 1st IEEE Workshop on Broadband converged Networks (BcN), held as part of IEEE/IFIP NOMS, Vancouver, Canada, April 7, 2006.

[20] TeleManagement Forum (Ed: J. Reilly), *Shared Information/Data (SID) Model - Business View Concepts, Principles, and Domains.* GB922, Ed. NGOSS R6.1, Document Version 6.1, November, 2005.

[21] TeleManagement Forum (Ed: J. Fleck), *NGOSS Architecture Technological Neutral Specification - Distribution Transparency Framework Services*. TMF053F, Ed. TNA 6.3, Document Version 1.0, January, 2004.

[22] J. Strassner, N. Agoulmine, E. Lehtihet, *FOCALE – An Autonomic Networking Architecture*. In the ITSSA Journal, 2007.

[23] http://www.bsi-global.com/en/Standards-and-Publications/Industry-Sectors/ICT/IT-service-management/ITIL-guides/

[24] http://www.zifa.com

[25] S. Balasubramaniam, K. Barrett, J.Strassner, W. Donnelly, S. van der Meer, *Bio-inspired Policy Based Management (bioPBM) for Autonomic Communication Systems*. In Proc. of 7th IEEE Workshop on Policies for Distributed Systems and Networks (Policy 2006).

[26] M.-J. Choi, H.-T. Ju, J. Won-Ki Hong, D.-S. Yun, *Towards Realization of Web services-based TSA from NGOSS TNA*. In Proc. of 6th IEEE International Workshop on IP Operations and Management (IPOM'06), Dublin, Ireland, October 23-25, 2006.

[27] B. Jennings, S. van der Meer, S. Balasubramaniam, D. Botvich, M. Ó Foghlú, W. Donnelly, J. Strassner, *Towards Autonomic Management of Communications Networks*. IEEE Communications Magazine, **45:10**, 2007, 112–121.

[28] http://www.manweek2006.org/mace/program.php

Acknowledgement

The authors wish to thank the members and contributors of the TMF Architecture and the Lifecycle & Methodology teams for their contributions, discussions and patience.

Part of this work has received support from the Science Foundation of Ireland under the Autonomic Management of Communications Networks and Services programme (grant no. 04/IN3/I404C).

Sven van der Meer
Waterford Institute of Technology
Telecommunications Software & System Group
Cork Road
Waterford
Ireland
e-mail: vdmeer@tssg.org

Joel Fleck
Hewlett-Packard Company
1001 Frontier Road, Suite 300
Bridgewater, NJ 08807
United States
e-mail: joel.fleck@hp.com

Martin Huddleston
QinetiQ
Applications & Services Group
Bldg A8 Rm 1004
Cody Technology Park
Ively Road, Farnborough
Hants, GU14 0LX
United Kingdom
e-mail: mehuddleston@qinetiq.com

Dave Raymer
Motorola Inc
Motorola Labs
1301 E Algonquin Road
Schaumburg IL 60196
United States
e-mail: david.raymer@motorola.com

John Strassner
Motorola Inc
Motorola Labs
1301 E Algonquin Road
Schaumburg IL 60196
United States
e-mail: john.strassner@motorola.com

Willie Donnelly
Waterford Institute of Technology
Telecommunications Software & System Group
Cork Road
Waterford
Ireland
e-mail: wdonelly@wit.ie

Whitestein Series in Software Agent Technologies, 27–42

A Telco Approach to Autonomic Infrastructure Management

José A. Lozano López, Juan M. González Muñoz and Julio Morilla Padial

Abstract. *"The next thing in technology is not just big but truly huge: the conquest of complexity"*, Andreas Kluth, The Economist [7]. The future of the Information Society draws a world where devices with processing capabilities are ubiquously connected to information sources. Due to the diversity of user devices, technologies, communication accesses, etc. operators do really need a technology able to manage such a complex environment giving a service end to end vision. In this sense, the set of technologies that arise from Autonomic Communication appears to be the cornerstone in order to manage the complexity of these new networks and services infrastructure. Competitive market is hard and telecommunication operators must face a double challenge, on one hand they must offer powerful services that must be managed, on the other hand they must address the reduction of service margins. Only really efficient operators will survive in such an environment. The I2OSS (Intelligence to Operational Support Systems) model is thought to provide operators with management infrastructures showing the advantages of Autonomic Communications. The reader will find in this article a technological framework for innovation in the field of Autonomic Communications. The I2OSS model establishes a methodological approach for innovation on OSS (Operation Support Systems) systems that support the management processes of a telco operator. No concrete or specific solutions will be found, but indications on how a telco operator must drive innovation to take advantage of "Autonomic Technologies".

1. Introduction

The future Information Society that we can imagine shows a world with individuals seamlessly connected to other individuals or machines. These connections will allow them to access and use pervasive computational resources. These environments will provide a quantum step in the communication and information services improving people's quality of live.

Without looking so far, nowadays personal communication environments present an increasing number of heterogeneous devices such as computers, PDAs (Personal Digital Assistant), webcams, home gateways, sensors, etc. that can be used from different places to access a myriad of very different services or applications. Building this scenario requires a new reliable, dynamic and secured communication infrastructure with highly distributed capacities.

Another characteristic of current environments is that some of the tasks required for the installation, configuration, and maintenance of applications and devices have to be carried out by users. These operations are not usually an easy task for non expert users. As a result, this situation becomes a handicap for the development of the Information Society.

The complexity of managing such infrastructure exceeds the capabilities of current OSS and is the main challenge that telecom industry is currently facing. Service Providers (SP) have to cope with all this complexity in order to have their customers enjoying useful and ergonomic services. SPs have to become a "Black Hole" managing all this complexity. Operators need to change their current management paradigms if they want to avoid collapsing under the operational weight of managing complexity.

In 2001, Paul Horn, IBM senior vice-president of research, in a keynote to the National Academy of Engineers at Harvard University, pointed out complexity as the barrier for IT (Information Technologies) industry's further evolution and he pointed out Autonomic Computing as a model to go beyond this barrier [8].

The term Autonomic Computing was chosen to make explicit the analogy with the autonomous nervous system. An autonomic computing system has the faculty of self-management driven by high level goals and users' objectives. In Telefónica I+D, we are extending the principles and concepts to telecommunication networks and services, as the Autonomic Communication initiative [1] proposes.

Communication networks and their management systems have to be able to be autonomous and seamless from users' point of view, they must react to context changes. Moreover, they have to dynamically incorporate new objectives and knowledge about business and operations.

In order to deploy fully autonomous communications, some technological breakthroughs are needed in the areas of systems cooperation, intelligence and knowledge.

This paper shows **I2OSS** (*Intelligence to OSS*) [5] a conceptual model developed in Telefónica I+D to deal with the requirements of these new communication infrastructure and services. This model is aligned with the Autonomic Communications initiative, as one of its main objectives is to build highly autonomous and intelligent telecommunication and information infrastructures.

I2OSS builds a loop between the communication infrastructures, the systems that manage them and the person who operates and exploits services. Information is fed into the loop from the resources and the environment to the higher levels

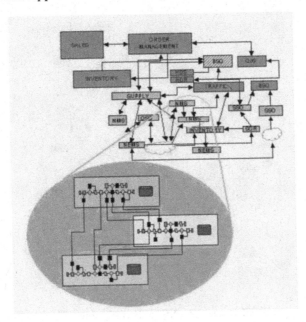

FIGURE 1. System to system connections.

and it is then fed back after processing to the resources to increase the automation level.

I2OSS establishes an innovation framework for telco operators to apply Autonomic Communications technologies. It highlights relevant points where autonomic technologies may be successfully applied to OSS.

2. The Mess of Systems

The geographical extension of the communication infrastructure and the amount of data to manage (users, equipment, transactions, etc.), have forced the telco companies to be big consumers of IT/IS (Information Technologies/Systems) from its beginnings. Automation of several and different tasks and software technologies being more and more feasible, have driven operators to an important increase on their number of systems. Today, the number of applications/systems already deployed in an incumbent operator can go beyond one thousand. Each end to end business process can be supported by more than one hundred systems.

In order to become more effective, higher automation levels were obtained by connecting systems. These connections were thought in a system-to-system basis with a particular solution for each case. As shown in Figure 1, systems present hard-coded implementations of pieces of business processes, producing rigid and inflexible structures. Changes in business processes propagate for a number of

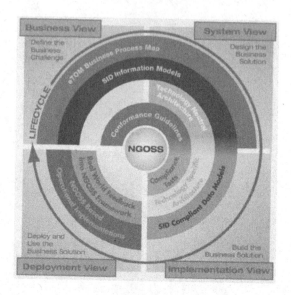

FIGURE 2. NGOSS overview.

systems and it takes months to effectively adapt all systems involved in those changes.

SOA (Service Oriented Architecture) [11] provides a solution for the lack of flexibility as it is supposed to solve the decoupling between the implementation of applications and the function they provide.

Focusing on network and services management systems, the application of SOA philosophy and concepts has been materialised by the TeleManagement Forum (TMF) in the NGOSS (New Generation Operation Software and Systems) [3] initiative. NGOSS is a comprehensive, integrated framework for developing, procuring and deploying operational and business support systems and software. It is available as a toolkit of industry-agreed specifications and guidelines that cover key business and technical areas. It is somehow the application of SOA concepts to the whole lifecycle of management systems.

NGOSS provides all the architectural concepts to apply SOA to the networks and services management area and defines the SID (Shared Information Data model) [12] to establish a common language between applications or systems.

Unfortunately, NGOSS is based on a static and deterministic view and so are current technologies involved in systems development. The management of a service based, highly dynamic infrastructure overflows the capacity of systems statically designed and built. When business requirements change, systems have no ability at all to evolve. For this issue an intelligent and autonomic model is mandatory.

Autonomic Communications initiative (AComm) [1] can help telco companies in facing these problems. AComm is a new research initiative grouping a

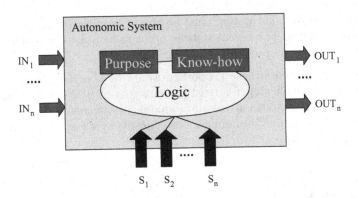

FIGURE 3. Autonomic System Description (Source: [4])

number of researchers from different scientific disciplines to evolve in the levels of self-management and self-control needed for the new generation communication infrastructure.

One of the main AComm objectives is to provide a new communication infrastructure with autonomous capacities. In this context, autonomous means the capability to work and react to changes in the environment without needing a centralised control, especially in unknown situations.

AComm provides more distributed and self-organising structures, relying on the behaviour of individual elements as part of a broader organisation. Evolvability and emergence behaviour are two key points in this area.

An important concept for AComm is that of Autonomic System (AS) stated in [4] as "a system that operates and serves its purpose by managing itself without external intervention even in case of environmental changes". Figure 3 presents a description of an autonomic system where a fundamental block of the AS is its capability to observe the external operational context, represented in the figure through S1 to Sn sensing inputs.

Another inherent block of an AS is the goal or purpose it serves, but also the know-how in order to achieve these objectives. Logic is the block responsible for making decisions to serve the system's purpose, but taking into account the observations of the context.

Following this definition of Autonomic System it is important to clarify those characteristics that make a system behave as an autonomic system. These properties are:

- **Automatic**: The system must be able to self control its internal functions and operations.
- **Adaptive**: An autonomic system must be able to change its operation or behaviour (i.e., its configuration, state and functions).

- **Aware**: An autonomic system must be able to monitor (sense) its operational context as well as its internal state in order to be able to assess if its current operation serves its purpose.

Any system that presents these properties would be classified as an Autonomic System. Then, the next step for trusted management of Next Generation Networks in order to accomplish a real and dynamic service management view is the application of Autonomic Communications concepts over a system framework like NGOSS.

3. The I2OSS Model

I2OSS is a conceptual model based on the principles and concepts of Autonomic Systems to bring autonomic characteristics to current management of telecommunication infrastructures. It applies to the global operators' infrastructure including all elements from network itself to management systems. Each layer implementation is distributed all over them.

Translating these concepts to telco companies' management models, technologies and systems must hide complexity. Workforce staff shouldn't be worried about specific technologies but business requirements instead: They in turn must worry about fulfilling customer expectations. It does not develop any new technology but it establishes the foundation to place a number of technologies and initiatives in this area.

3.1. Description

The I2OSS model is currently in development and establishes three different abstraction levels to implement the concepts and principles of autonomic communications over the operator's telecommunications infrastructure. Higher layers correspond with higher levels of abstraction. These three layers are depicted in Figure 4.

- **Intelligence Layer**: In this layer business goals are defined and decisions are taken in order to achieve them. Information is fed from other layers and then corrective actions are sent back to the infrastructure. In order to control the infrastructure, control mechanisms as policy based management are taken into account.
- **Information Layer**: The main objective of this layer is the semantic management of the information that is generated in the infrastructure. Its main contribution is that it provides a common vehicle for the communication between entities. Knowledge management is somehow between the intelligence and the information layer. It can be seen that the information layer conceptualises knowledge while the intelligence layer is responsible for materialising it.

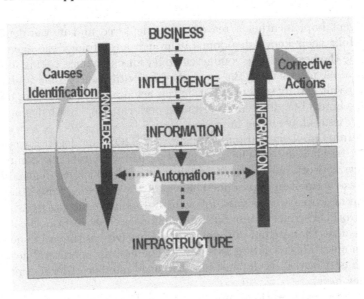

FIGURE 4. I2OSS Model.

- **Automation Layer**: This is the bottom layer, where automatic infrastructure operation is done. Modern network elements and resources include more processing capacities than ever and they can assume more tasks.

An important aspect of this model is that it shows a dynamic behaviour in several senses: relations between different levels are dynamic and so is their evolution.

Information flows from the infrastructure to the intelligence layer. Knowledge flows in the opposite direction from intelligence level to the infrastructure. Intelligence establishes rules that are materialised throughout the different abstraction levels in their way towards physical resources.

As it has been stated, intelligence is distributed over different entities, and although each entity shows an autonomous behaviour, it is necessary to model the relations among them. The true goal is to accomplish business objectives as the sum of individual autonomic entities goals. This aspect is multidisciplinary as it is necessary to investigate from information technologies to social intelligence issues. These models resolve complex problems based on the aggregated effect of minor entities with few capacities.

Technologies, that may be applied to each layer in order to implement the definitions stated above, are at this moment quite immature but promising instead. For example, we have checked the potential use of inference engines on residential gateways [6]. The goal was to show a scalable solution for customers' network management.

An agent implementing a reasoner was deployed and it was demonstrated that describing their knowledge domain through an ontology the agent was able to determine the best suitable commercial offer for customers, taking into account the technical environment of the user together with their profile and operators' rules. As a result, a fully scalable solution was achieved. But for future use of these technologies in real life, different aspects like performance, knowledge management and security issues have to be improved.

Concerning implementation aspects for the I2OSS model, two ways of doing things must be highlighted. The first one is to achieve autonomic communication through new network and systems design; this is obviously the cleanest method to give infrastructures an autonomic behaviour. Nevertheless, this situation where systems and networks are designed following autonomic concepts is not applicable to the real world. The most common situation for telcos starts from the necessity to manage resources that are not autonomic by themselves, as they were not designed to be autonomic. This situation is the most common in operators' networks and services infrastructures, see for instance the huge number of legacy management systems they have.

In order to give traditional infrastructures the desired autonomic behaviour, *Agent* based technologies are the key to overcome this handicap. Agents will play different roles inside the I2OSS model, that is to say that an agent may be specialised in one or more layers. An agent may implement an inference engine at the intelligence layer or may be dedicated to execute actions at the infrastructure layer. Agents are fully scalable and suitable for distributed infrastructures.

3.1.1. Intelligence Layer. At this layer, the first approach to give intelligence to systems is enabling them to make inferences based on knowledge. That is to say they must be able to execute some kind of reasoning. Technically one solution for this approach is the implementation of inference engines. These engines are able to analyse information described using ontologies, and they also take into account behaviour rules defined. If the information or any rule changes, the output of the inference engine will also adapt to the new scenario.

Technical solutions to add intelligence to systems and networks are mainly based on software agents and autonomous elements which will exploit these capabilities in order to globally behave as planned for the autonomic system or systems.

Note that for the scope of this paper an agent is a tool to transform non autonomic elements into autonomic elements. From now on, an Autonomic Element (AE) is any entity showing autonomic capacities.

3.1.2. Knowledge and Information Layer. This layer is one of the keys for an infrastructure to show intelligent capabilities. Knowledge not only stands for information models that, of course, are necessary at this stage, but it is responsible for enabling machines to read and understand information and data. In traditional models, information is thought to be read and understood by humans, for example a web page that is written in HTML is designed for humans to read it. Machines are not able to understand it, because the semantic of the web page is implicit in readers' knowledge. Therefore a new model showing machine-readable character-istics is needed. Autonomic systems need to have the power of understanding data and obtaining new relations without having been specifically programmed for this task.

In order to support the intelligence required for an autonomic system two elements are necessary, on one hand the reasoning capability, on the other hand in order to apply knowledge it is necessary for the autonomic system to have a formal description for knowledge. Reasoning capabilities are the main focus of the intelligence layer. The Information Layer will worry about machine readable domain descriptions.

Ontology is one of the main modelling techniques used today to represent information that can be understood by machines. That is to say, it is a way to represent the knowledge that a machine requires to interpret information and data.

The definition of ontology has been stated by the OMG (Open Management Group) as follows:

"Ontology, defines the common terms and concepts (meaning) used to describe and represent an area of knowledge. An ontology can range from Taxonomy (knowl-edge with minimal hierarchy or a parent/child structure) to a Thesaurus (words and synonyms) to a Conceptual Model (with more complex knowledge) to a Log-ical Theory (with very rich, complex, consistent and meaningful knowledge). A well-formed ontology is one that is expressed in a well-defined syntax that has a well-defined machine interpretation consistent with the above definition"

Using ontology at this layer allows taking advantage from technologies originally developed for the Semantic Web area. Ontology definition using a standard "de facto" language like OWL (Ontology Web Language) is suitable for the needs of OSS systems. It is also recommended to complement this language with SWRL (Semantic Web Rule Language) to describe explicit rules.

Concerning the information model, operators' information needs are very well described by the Shared Information and Data Model (SID) that has been developed at the TMF. As the SID is fully compliant with operators processes described by eTOM [14], it is a quite good initiative to drive business goals. A successfully Information Layer must combine the SID information model with the potential of ontologies. This fact will give to the intelligence layer an end to end business view of the operator.

3.1.3. Infrastructure Automation Layer. This layer is the place where well known infrastructure lies. Autonomic networking technologies and protocols are found dealing with self-management for physical infrastructures, responding to different issues as for example QoS (Quality of Service), routing or security policies. For logical infrastructures, technologies should accomplish a decoupling and service oriented view. In addition, some mechanisms must be applied in order to simplify

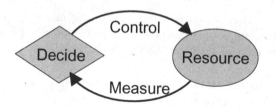

FIGURE 5. Control loop.

service orchestration. In this sense the application of SOA concepts is the key towards automation of infrastructure.

It is also very important to consider policies at this stage, as the I2OSS infrastructure automation is controlled through the use of policies so its resources behaviour must be policy driven. Policies are the mechanisms that will provide the operator the possibility to change the infrastructure behaviour without having specific knowledge about the resources inside.

Policy based management is nowadays statically implemented, and it really facilitates the management and operation over resources. High level policies defined in a centralised point are distributed and translated to specific rules for specific resources. Although this mechanism facilitates operation, it is not suitable for autonomic communication, as it will not be able to overcome context changes or eventual conflicts between resources that have not been explicitly programmed to deal with particular situations.

3.2. Coordination aspects

The orchestration of the different functionalities located at each layer is based on Control Theory. The next subchapters detail the concepts and requirements that are needed in order to effectively coordinate functionality from the layers described above. The solution proposed in this paper is about creating control loops that act as managers of resources through monitoring, analysis and taking action based on a set of policies.

Figure 5 shows the operation of a closed control loop. There are two extremes or nodes, the monitored resource and the monitoring intelligence. The monitoring intelligence node senses or receives information from the resource. Depending on these parameters, it processes the information and decides orders and controls back the resource.

These control loops can communicate with each other in a peer-to-peer context and with higher-level managers. For example, a database system needs to work with the server, storage subsystem, storage management software, the web server and other system elements to achieve a self managing IT environment. This communication may be achieved through the information layer, that is to say knowledge is updated. For example if a control loop changes something in the

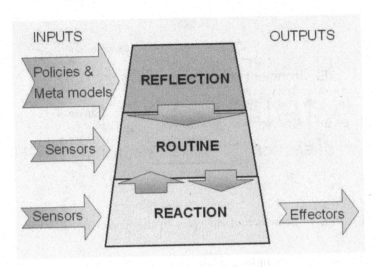

FIGURE 6. Levels of Knowledge.

domain, the ontology must reflect this change. In this way all entities involved in other control loops are aware of the domain state.

There are three levels that will interact in order to obtain the desired elements automaticity. These levels from top to bottom are called:

- Reflection
- Routine
- Reaction

The reaction level is the lowest, it has no learning abilities, its purpose is to give immediate response to state information coming from sensory systems and from the routine level. Effectors are only controlled at this level so the element can react to stimulus from sensors always overseen by the routine instructions.

Routine is in the middle, where routine evaluation and planning behaviour take place, it receives inputs from sensors and from the other two levels.

Reflection is the top level, which receives no input from sensors and has no motor output. Policies and meta-models are introduced at this level so this meta-process can deliberate about itself. It considers the overall current behaviour, current environment and its experiences to learn new strategies that are sent to the lower level. Figure 6 shows the relationship among these three levels.

To make a system based on this architecture we need to work mainly with AI (Artificial Intelligence) and engineering techniques. Essentially the lowest level will be designed using engineering technologies to make sensors and effectors fulfil their function. The reflection level may use AI techniques to consider the system's behaviour in order to learn new strategies. As for the routine level, it would use a mixture of both.

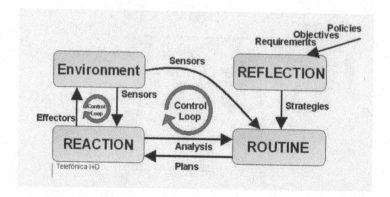

FIGURE 7. Sensors and Effectors in the Knowledge lifecycle.

The AI techniques include machine learning, Tabu search, statistical reasoning and clustering analysis. They can also include soft techniques such as neural networks, fuzzy logic, probabilistic reasoning incorporating Bayesian networks, etc.

From the engineering domain we can use cybernetics optimisation techniques, fault diagnosis techniques, feedback control, planning techniques, etc.

3.2.1. Sensors and effectors. The brain of the I2OSS is composed by the intelligence and the information layer, together they accomplish decision and knowledge generation. However, just as a body is nothing without its ears, eyes or hands, an autonomic element is nothing without its sensors and effectors.

Sensors extract the necessary information from the environment for a later monitoring and analysis. On the other hand, effectors execute the system's plans. Sensors and effectors are the media by which an AE interacts with the environment.

According to the knowledge cycle proposed in the previous section, Figure 7 shows the relationship among Reaction, Routine, Reflection and the Environment, and where sensors and effectors take part.

New methods and management tools will cope with this complexity in order to react properly to stimulus from sensors. An adaptive behaviour will enable the system to react to changes and to learn and evolve.

3.3. An evolving model

The I2OSS model does not start from scratch, it is thought to take current systems and architectures as a starting point, which is the real situation of telecommunication operators. Different levels are built in an independent way, and relations among them are built as knowledge management is developed.

This idea, with a more limited scope, has already been applied by Telefónica I+D in the OMEGA project [13]. As a result, problem management for the switching network has been automated. This success is based in an organisational model and a clear based method. The working method allows to capture knowledge from

O+M (Operation and Maintenance) technicians and to implement it on a management agent whose behaviour is driven by this knowledge. Agents monitor events and act in the same way human technicians would act.

I2OSS has more ambitious challenges, as on one hand it captures knowledge and on the other hand provides the necessary infrastructure to apply it over different technological platforms in order to achieve business goals.

So I2OSS model takes into account that it is impossible for a telco operator to make all its systems to become autonomic in just one step. IBM Autonomic computing initiative was also thought as an evolutionary process from the manual management towards the autonomy. This autonomous final stage can be reached in five levels, which are defined in [10] as follows:

1. **Basic level**. A starting point of IT environment. Each infrastructure element is managed independently by IT professionals who set it up, monitor it and eventually replace it.
2. **Managed level**. Systems management technologies can be used to collect information from disparate systems onto fewer consoles, reducing the time it takes for the administrator to collect and synthesise information as the IT environment becomes more complex.
3. **Predictive level**. New technologies are introduced to provide correlation among several infrastructure elements. These elements can begin to recognise patterns, predict the optimal configuration and provide advice on what course of action the administrator should take.
4. **Adaptive level**. As these technologies improve and as people become more comfortable with the advice and predictive power of these systems, we can progress to the adaptive level, where the systems themselves can automatically take the right actions based on the information that is available to them and the knowledge of what is happening in the system.
5. **Autonomic level**. The IT infrastructure operation is governed by business policies and objectives. Users interact with the autonomic technology to monitor the business processes, alter the objectives, or both.

These five levels are not mandatory, they are just theoretical, so operators can skip some levels provided they reach the aimed one. More over, the speed in the adoption and implementation of autonomous systems depends on each enterprise.

3.4. Non technical aspects for auto-coordination.

Another important aspect of autonomous systems is that related to the auto-coordination of autonomic systems and elements. This is a truly big challenge. Not only principles about networks and communications must be applied, principles from other disciples as economical, societal relations or intelligence science are also needed. Mechanisms for role based management, evolution mechanisms (as genetic algorithms), neural learning, etc. are key technologies for autonomic communication development.

In this sense, using economics theories to organise the telco infrastructures is becoming a useful tool as they can provide models for resources contributing to the same objectives with a cost efficient operation structure.

In fact, the behaviour of several autonomous systems matches perfectly with the definition of market and it can be viewed as the management of limited resources to accomplish a given objective. Using this approach the network and the systems can flexibly adapt to the business goals.

There are three main topics that must be defined in order to get an optimal result and to incorporate the business goals and the consumer expectations in the network behaviour and management:

1. Legal environment of the market: We have to define the policies and rules needed for a correct working of the trading between agents and final users to influence the right behaviour of the agent to reach the optimal equilibrium, giving to agents an incentive to truthfully reveal how much they value resources. In this case, we need a market creator to assure the right trading of services between ASs.
2. Demand Function: A utility function is use to model consumer preferences in our application to the I2OSS. Through the estimation of indirect utilities function of users using contingent valuation techniques we can measure the economic values of new services or services portfolios to be introduced in the future. This function aligns AS services demand with business goals such as QoS, end services revenues, customer expectations and cost efficiency.
3. Offer Function: A production function must reflect the capacity, the cost of provided services and the facility to use resources to create other services.

The use of economic theory could help to provide to the AS the criteria to organise themselves.

4. Conclusions

The connected society requires pervasive computing. It draws a world where processing power, storage capacity and communications are used by users without worrying about specific technologies.

The required infrastructure to provide this functionality must be highly adaptive and autonomous. The resources that compose the infrastructure should have dynamic relations among them. So complexity management will require new management systems generation.

This paper has described a conceptual model, I2OSS systems based may be used in order to include intelligence in operator's processes. The I2OSS model is based on Telefónica I+D experience on management systems, and is employed to automatically capture knowledge and to apply it to network and services management. Its final goal is getting reasonable levels of automation, flexibility and dynamics in business processes.

Furthermore, it can be implemented starting from SOA architecture with a centralised processes control entity and evolving it towards a more distributed architecture.

Another aspect stated in this paper is that related to information as the basis for knowledge. Semantics and ontology technologies allow a universal access to information from an operational point of view as well as for a business view.

It is important to point out that almost all these technologies are in an initial stage and present an important lack of maturity in order to get the desirable stability. Anyway we believe they are the future.

In order to apply Autonomic Communications to current operators' infrastructure, it must be considered that current networks and systems are not autonomic at all and this is the starting point. As it is impossible to start from scratch, current infrastructures must evolve to show an autonomic behaviour. In the short term software agents technologies are the key to add autonomic functionality to current networks and systems, but in the mid and long term, a new framework must be developed to build a new generation of infrastructures (networks and systems) with inherent autonomic capabilities.

5. Glossary

AC/AComm	Autonomic Communications
AE	Autonomic Element
AI	Artificial Intelligence
AS	Autonomic Systems
I2OSS	Intelligence to OSS
IS	Information Systems
IT	Information Technologies
NGOSS	Next Generation OSS
O+M	Operation and Maintenance
OMG	Object Management Group
OSS	Operational Support Systems
OWL	Ontology Web Language
QoS	Quality of Service
SID	Shared Information Data
SOA	Service Oriented Architecture
SP	Service Provider
SWLR	Semantic Web Rule Language
TMF	TeleManagement Forum

References

[1] M. Smirnov, *Autonomic Communication, Research Agenda for a New Communication Paradigm*. Fraunhofer Fokus, White Paper, November 2004.

[2] J. C. Strassner, N. Agoulmine, E. Lehtihet, *FOCALE - A Novel Autonomic Networking Architecture*.

[3] TeleManagement Forum, *The NGOSS Technology Neutral Architecture*. TMF 053, Release 6.0, Nov 2005.

[4] S. Schmid, M. Sifalakis, D. Hutchison, *Towards Autonomic Networks*. In proceedings of 3rd Annual Conference on Autonomic Networking, Autonomic Communication Workshop (IFIP AN/WAC), Paris, France, September 25-29, 2006.

[5] J. A. Lozano, J. M. González, F. J. Jariego, *OSS inteligentes: Conquistando la complejidad de los servicios de la informacin y las comunicaciones*. Revista Comunicaciones I+D, **38**, Abril 2006.

[6] J. M. González, J. A. Lozano, J. E. López de Vergara, V. A. Villagrá, *Self-adapted Service Offering for Residential Environments*. Proceedings of 1st IEEE Workshop on Autonomic Communications and Network Management, ACNM07 Munich.

[7] A. Kluth, *Make it simple*. Published in The Economist, October 28th, 2004.

[8] P. Horn (Senior Vice-president), *Autonomic Computing: IBM's perspective on the State of Information Technology*. IBM Research.

[9] K. J. Arrow, G. Debreu, *Existence of an equilibrium for a competitive economy*. Econometrica **22**, 1954, 265–290.

[10] IBM, *An architectural blueprint for autonomic computing*. Whitepaper, June 2005.

[11] OASIS, *Service Oriented Architecture Reference Model*.

[12] TeleManagement Forum, *Shared Information/Data Model: Concepts, Principles and Domains*. GB 922. Release 4.5, November 2004.

[13] G. Redondo, L. De Miguel, J. Valderrama, *OMEGA: Innovacin Tecnolgica y Organizacional*. Revista Comunicaciones I+D, **35**, Marzo 2005.

[14] TeleManagement Forum, *Enhanced Telecom Operation Map (eTOM): The Business Process Framework*. GB 921. Release 6.0, November 2005.

José A. Lozano López, Juan M. González Muñoz and Julio Morilla Padial
"Autonomic Communications Division", Telefónica I+D
C/ Emilio Vargas 6
28043 Madrid Spain
e-mail: {jal,jmgm,morilla}@tid.es

Whitestein Series in Software Agent Technologies, 43–62
© 2007 Birkhäuser Verlag Basel/Switzerland

Modelling Behaviour and Distribution for the Management of Next Generation Networks

C. Fahy, M. Ponce de Leon, S. van der Meer, R. Marin, J. Vivero, J. Serrat, N. Georgalas, P. Leitner, S.Collins and B. Baesjou

Abstract. Current network management systems have been impeded by a scarcity of open standards for interoperable management solutions. Information models have made progress in promoting interoperability of traditional "centralised" networks but have still to significantly address the proliferation of next generation networks such as autonomic networks. Such networks impose challenges which include distributed self-control and self-management. The goal of the Madeira project was to utilise novel technologies and methodologies, based on an underlying P2P paradigm, for a logically meshed, distributed Network Management System (NMS) that facilitates dynamic behaviour of transient network elements. In this paper, we describe a solution for a meta-model that attempts to capture the key concepts behind the task of network management of a mesh network. A case-study focusing on the fault management of such a network will be presented with the purpose of verifying the applicability of such a meta-model.

Keywords. meta-model, autonomic networks, management, distribution.

1. Introduction

Voice and data communication networks are continually evolving to contain very large numbers of nodes, involving network of networks (with more operators, equipment vendors and owners) and with larger technological diversity and heterogeneity of network elements [1].

This places a requirement on network management systems to be able to deliver more adaptive and simpler solutions for next generation networks such as autonomic networks [2]. Currently deployed management systems are constrained by static architectures that insufficiently allow for flexibility and have rigid interoperability standards [3]. As networks become larger and more heterogeneous, and exhibit increasingly dynamic behaviour, this historical approach becomes inadequate.

Key new capabilities required for network management of such systems include: self configuration within a network [4], of intra- and inter-network collaboration for management tasks [5], more adaptability to service requirements and support for varying network deployment scenarios.

The Peer-to-Peer (P2P) network paradigm [6] demonstrates some of the challenging management problems that can be associated with Next Generation Networks [9]. Exploiting P2P characteristics in order to achieve self-organisation, symmetric communications and distributed control [8, 9], can lead to a more adaptive network control for today's transient network element systems.

Modelling, and specifically meta-modelling, can be used to understand and process the complexity of requirements for network management. Our main focus is to realise network management capabilities for P2P networks and ultimately NGNs, thus the following question characterises our modelling research focus: "What are the additional information and behaviour that need to be embedded into, or built on top of, network management protocols to help produce a distributed, communications management system without a static hierarchy?"

Our research work is done through the European Celtic initiative (a Eurescom cluster) [10] project Madeira [11] that aims to provide novel technologies and methodologies for a logically meshed, distributed Network Management System (NMS) that facilitates dynamic behaviour of transient network elements. The goals of the project are to:

- Specify an architecture for a distributed NMS, which will allow a high degree of inter-working between the management systems of various network domains.
- Develop a logically meshed Peer-to-Peer network management computing environment which can provide an advanced computing reference framework to support management operations that are massively distributed in nature, across dynamically forming networks, and to facilitate distributed management application development.
- Provide a means of rapidly and efficiently describing and programming management operations that form network management applications through new modelling techniques.
- Explore through a case study the new relationship between Configuration and Fault management for transient dynamic network elements, with concerns such as, how to differentiate between fault and normal network behaviour, being addressed.

The project takes a scenario based user-centric approach to requirements definition and emphasises the services that the system elements shall offer from the perspective of the users (i.e., Operators and Application Developers).

The core of this project is the definition of the architectural requirements, computing and communications platform and meta-data modelling. While these three areas are intrinsically linked, this paper focuses specifically on the definition of the network management meta-data and behaviour for the network management of large-scale Peer-to-Peer networks.

2. State of the Art

A model is used to describe the characteristics, functions, structure and/or behaviour of a system or of one of its parts. Due to the heterogeneity of currently available network management models such as SNMP [12], CMIP [13], DMI [14] or WBEM [15, 16], the necessity for mechanisms which enable interoperability between the various models has grown. While mapping across models has been a popular solution, it has proved more desirable to take the problem to a higher level and derive a set of concepts or semantics that has meaning to all models. Such a set of semantics is known as a meta-model and is used to define what can be expressed in a valid model. A meta-model expresses the necessary semantics once, thus enabling the transformation towards different technologies.

UML is the original meta-model and as such defines a common and unified set of semantics to be defined that will be used in all models. The Meta Object Facility (MOF) [17] is a 4-layer meta-modelling architecture used to define UML. The top layer is a meta-meta model which defines the structure and semantics of the meta-model in the next layer. The meta-model is then used to build the model layer and the last layer represents the actual data which needs to be described.

The NGOSS meta-model is an extension of the UML meta-model. It was extended in order to include NGOSS specific concepts and information [18]. This meta-model is used as the foundation for all the NGOSS Technology Neutral Architecture (TNA) specifications, for example: Shared Information and Data Model (SID). This meta-model introduces concepts such as: NGOSS Contract - represents interoperability, NGOSS component - a standardised way of packaging NGOSS functionality, NGOSS Shared Information - information that is designed to be shared amongst NGOSS components, NGOSS policy - used to specify behaviour of an NGOSS element and NGOSS interaction - specifies how NGOSS elements interact. While many of these concepts provide a means to represent behaviour and interoperability within a management system, their definitions are not rich or descriptive enough to capture the distributed behaviour of a self-managing system such as a P2P or Autonomic System.

The Model of Primitives (MoP) [19] is used to model complex structures by providing a set of basic but expressively rich semantic constructs. If we were to map MoP to the MOF layered architecture, it would be present at the meta-meta model level or M3 layer. MoP includes policies and events in its definition thus incorporating behaviour from the highest level of modelling.

The Middleware and Application Management Architecture (MAMA) [20] is a body of work that attempted to bring together renowned approaches to middleware and management to create a unified architecture. MAMA adopted concepts from such prominent distributed processing and management standards/ approaches as the RM-ODP [21], TINA-CMC [22], CORBA [23] and DMTF-CIM [24]. MAMA provides a meta-schema that defines basic specification elements that has relevance for any application model.

3. Madeira Architecture

Before continuing, we take a moment to briefly describe the Madeira architecture solution that is used to build our P2P, distributed network management system. The defined Madeira Architecture attempts to encapsulate a reference framework that provides management operations that support the distributed and transient nature of the network's elements. A logically meshed Peer-to-Peer network management environment is developed to achieve this.

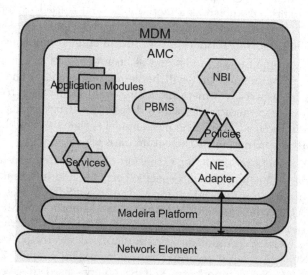

FIGURE 1. Madeira Distributed Management System

The Madeira Distributed Management System (MDM) is briefly comprised of two separate but co-operating entities - application functionality encapsulated within the container that is the AMC and the platform which provides all the generic functionality required to support the network management tasks of the AMC. The collective solution that makes up the MDM can be deployed on individual Network Elements (NEs) providing each NE with self-management capabilities whilst enabling a distribution of information and management functionality across the network of NEs.

Adaptive Management Components (AMCs) are P2P network management entities that support Madeira management functions. An AMC is responsible for the network management functionality of a peer in a P2P network and contributes to the collective distributed NMS functionality.

As can be seen in Figure 1, the AMC is structured into components - the application modules, the AMC specific services, the Policy Based Management System (PBMS) and policies which define the performance of the System, the Northbound Interface (NBI) service which provides an interface from the AMC towards an external OSS and the Network Element (NE) Adapter service which provides an interface towards the specific technology of the network element below.

The Madeira Platform provides a middleware consisting of P2P generic services which support the AMC functions and communication - lifecycle service, code distribution service, grouping service, directory service, communication service and persistency service.

The "applications" refer specifically to the Configuration Management and Fault Management application. These applications use the AMC Services and the Madeira Platform Services to form an overlay network of AMCs to create the intended management application logic. Such a configuration of AMCs residing on each NE is shown in Figure 2.

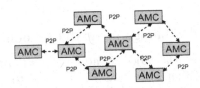

FIGURE 2. Overlay Management Network using AMCs

The management network is partitioned by the grouping service into logical clusters which then elect cluster heads (super peers). This logical hierarchical overlay network enables scalability for management of large-scale meshed wireless networks, and facilitates both self-healing, and aggregation and correlation of management data, so that the data reported to the operator is minimised (critical in such large networks).

3.1. Applying the Model Driven Approach

Considering the functionality split between AMC and platform in an actual instance of the Madeira Management System, the AMC covers the application specific parts for a particular scenario, whereas the platform provides all the generic functionality required for the network management tasks in the P2P environment.

This separation within the system ensures the feasibility of a model driven approach, using OMG's Model Driven Architecture (MDA) principles [25], for AMC development in order to adapt to changing scenarios and requirements in a very efficient way. MDA supports model-driven engineering of software systems - functionality can be firstly modelled as a Platform Independent Model (PIM) using an appropriate meta-language. This PIM can then be transformed into one or more Platform Specific Models (PSM).

The application of MDA can solve the problem of heterogeneity introduced by the different network technologies and applied standards as it allows the specification of shared / distributed behaviour (logic) and state (data) to happen once in a technology-neutral way. The transformation into numerous technology-specific formats can then be done automatically or semi-automatically as required by the various network elements and platforms. Particularly, we are using MDA in a twofold way in Madeira. First, we model information, which can then be transformed to different data models supported by the elements it will be deployed on. Second, we model NM application logic and transform it to formats that can be executed on different platforms.

The AMC and related applications are only bound directly to a software platform middleware and Network Management technology when a transformation has been made from a Platform Independent Model to the Platform Specific Model. The overall Model Driven Approach taken in Madeira is described in full detail in [26].

4. Proposed Network Management Meta-model

By defining a meta-model, the objective addressed is the provision of a minimum as well as generic set of semantics that will be used to build the model for our network management application or AMC for Madeira.

A Language driven development [27] approach was applied here. This involved the application of MDA [24] technologies to attempt to generate a semantically rich language and tool that targets the specific modelling requirements of distributed network management system. Modelling languages aim to provide rich abstractions that are aligned more closely with domain concepts used by the developer. Language driven development would make it possible to integrate multiple languages and to define semantically rich modelling capabilities at a high level of abstraction. In this way, traditionally informal modelling languages like UML can be made precise and semantically more useful.

4.1. Meta-Model Requirements

4.1.1. Generic Meta-Data Model Requirements.
A number of initial, generic requirements need to be fulfilled by the meta-model before it can be deployed:

- The meta-model should be a general framework that enables the engineer to represent general network management information;
- It should be expressed in a language that is close to the one the network management engineer actually uses in his/her daily tasks;

- It should be independent of any technology-specific (platform specific) aspects;
- It should allow network engineers to easily write and understand network management concepts while at the same time providing a mapping to formal software logics i.e., it should be formal and expressive enough to allow translation into a machine-readable format;
- It should allow network engineers to easily write and understand network management concepts while at the same time providing a mapping to formal software logics i.e., it should be formal and expressive enough to allow translation into a machine-readable format;
- The meta-model should be agile enough in order to grant a simple maintenance process;
- The meta-model should be stable enough such that changes in it will not cause instabilities in the lower level implementations. The meta-model should contain enough abstract primitives such that it is possible to handle any network management scenario.

4.1.2. P2P Network Management Meta-Data Model Requirements. The Madeira architecture described in section 3 illustrates the structure of the P2P network management entity (AMC) and its containment within the distributed management system and the platform on which it runs. This P2P network management architecture places requirements on the network management entity such as:

- Social Communication
  The AMC shall have the ability to interact with its peers. It will be aware of other services, applications and resources and can advertise functionality to other services.
- Distributed Application
  The AMC shall have the ability to partake in distributed applications.
- AMC Auditability
  The AMC shall have the ability to record its activities for the purposes of later auditing.
- Information Handling
  The AMC shall have the ability to receive, interpret and send information from and to external entities.
- Network Management Intelligence
  The AMC shall exhibit a degree of network management intelligence, such as: goals and reasoning of network administration and maintenance operations.

These requirements are the building blocks for the AMCs to be able to create a distributed NMS with the following critical characteristics identified by [8, 9]: 1) self-organisation, 2) symmetric communication and 3) distributed control.

4.2. Developing the meta-model

In creating this Madeira meta-model, two bodies of work were brought together and adapted to meet the requirements for a meta-model that could describe the network management of a mesh network. These bodies of work are described below.

4.2.1. The Model of Object Primitives.

The MoP [19] is used to model complex structures by providing a set of basic but expressively rich semantic constructs. MoP is characterised by its inclusion of policies and events in its definition thus incorporating behaviour from the highest level of modelling.

As the name would suggest, MoP is made up of primitives the most basic of which is the MoPPrimitive. This is not directly used but it has the role of a "parent" primitive from which all the other primitives inherit. The following primitives make up the MoP:

- State Primitive: used for modelling a state.
- Behaviour Primitive: used to describe the behaviour of an entity, similar to methods of a class. It can also be used to define the specific delivery mechanism of the behaviour.
- Collection Primitive: defines aggregations/groups of primitives. This enables more complex structures to be created.
- Relationship Primitive: models a binary association between two primitives.
- Constraint primitive: used to place restrictions on MoP primitives.
- Policy primitive: a set of rules which express some behaviour.
- Event primitive: invoked by a change. If an event of interest is triggered, the policies will be activated to deliver a particular behaviour.

4.2.2. Middleware and Management Architecture (MAMA).

The Middleware and application management architecture (MAMA) [20] is a body of work that attempted to bring together renowned approaches to middleware and management to create a unified architecture. MAMA adopted concepts from such prominent distributed processing and management standards/approaches as the RM-ODP [21], TINA-CMC [22], CORBA [23] and DMTF-CIM [24]. MAMA provides a meta-schema that defines basic specification elements that have relevance for any application model.

The MAMA meta-model is expressed in the form of a meta schema in the Unified Modelling Language (UML). The Meta Model defines six main elements: module, object, interface, attribute, action and parameter. Two additional elements include type definitions and qualifiers.

An object comes from the definition of a computational object of the Telecommunication Information Networking Architecture [21]. A module can be used to collect zero or more objects or other modules. The interface of an object groups operations, which are called actions and any number of attributes.

Type definitions are used to extend the basic data types by adding new types. Qualifiers allow the representation of information that is specific to the problem that is under analysis.

4.2.3. The Madeira Meta-Model Solution. MDA specifies the use of the 4-layered MOF architecture as described in Section 2. The MoP serves as the M3 layer while MAMA serves as the basis for the M2 (or meta-model) layer. Then by constraining the MAMA meta-model with MoP, we are inherently introducing distributed behaviour into the meta-model.

As a Network Management solution, it is important that the model represents the static and dynamic behaviour associated with the peer and network distribution requirements associated with an AMC.

There are two main types of distribution that are considered in the context of Madeira:

1. Peer Distribution - This relates to the communication between distributed peers. This is handled by the introduction of behaviour.
2. Application Distribution - This is the overall distribution of a network management application across the network.

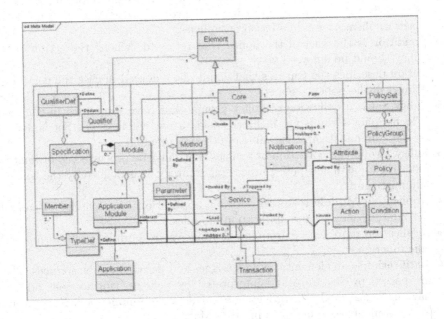

FIGURE 3. Madeira Meta Model

Figure 3 depicts the meta-model that can be used to describe the Madeira network management entity. The following provides a detailed description of its components and their functions:

- **Element** is an abstract class that is not to be instantiated. It's the necessary root of all the components of the language.

- A **Module** groups zero or more services, application modules and one core. It may also be "composed" of other modules. An example of the use of a module would be the representation of the AMC in the model.
- **Core** is the primary component of a Module and, based on notification and policies, orchestrates the services and applications to facilitate network management.
- **Services** are effectively toolboxes. They provide certain functionality required for the Core to manage. The Core invokes services.
- **Method** refers to method calls between services and the core. This replaces "action" from the MAMA meta-model.
- **Policy** is an entity that dictates one or more actions to be executed when zero or more conditions are met. It facilitates the dynamic and adaptive behaviour of the whole Madeira system. In particular, each AMC might be governed by policies and these policies are in charge of defining roles and responsibilities of each AMC within the network taking into account the capabilities of each AMC, the physical network and the needs of management applications. Moreover, each AMC can obtain a new specific role by dynamically downloading a new application module based on policies.
- **Condition** is the part of the policy that refers to "when" the action of the policy should be enforced.
- **Action** describes the features that should be enforced in case the conditions defined in the policy are met.
- **Policy Set** is a collection of policy groups that have something in common. Which Policy Groups are included inside a particular Policy Set is defined by the administrator at the time of creating them.
- **Policy Group** is a collection of policies that must be processed co-ordinately, following a particular policy processing strategy. It is defined by the network administrator.
- **Application Module** refers to portable application functionality. A module will contain one or more application modules.
- **Application** represents a grouping of applications modules to perform a specific function.
- **Notification** is an element that can trigger and be triggered by services. They are parsed by the Core and interpreted by policies. Notifications add behaviour to the model by providing a means of communication. Notifications are equivalent to events introduced in MoP.
- **Transactions** are records of some action that occurred.

5. Madeira Use Case Scenario

The scenario selected looks at the challenging area of self-management of wireless meshed networks, identifying a number of management tasks to be implemented using the meta-data model management approach advocated in this paper.

The scope of the scenario addresses two key areas of network management - configuration management and fault management. These chosen areas serve to highlight a set of "self-*" properties that are associated with autonomic systems - self-organisation, self-protection and of course, self-management.

The setting for the scenario is a large exhibition space where network connectivity is required by each exhibit. Wired connectivity is provided to machines at the periphery of the space while other machines must rely on wireless connectivity provided by the peripheral machines. It is important to point out that the management approach being championed in this scenario "a Wireless-LAN network" is a metaphor for any carrier-grade wireless or wired network.

The configuration management application handles the deployment and formation of the wireless mesh network, its re-configuration in case of fault or movement of network elements and it provides topology information northbound towards an external OSS for the purposes of topology visualisation.

The fault management application deals with fault alarms reported from a physical or logical (within Madeira) source. It performs alarm formatting, alarm correlation, analysis, enrichment functions when multiple alarms are generated for the same fault and it reports appropriate alarms northbound towards the OSS.

6. The Resulting Models

In validating our meta-model solution from Section 4, a model is built to describe our P2P network management entity or AMC that could be used to realise the network management scenario in Section 5. Once we describe this, we go on to present a case study of the meta-model's use in the fault management application.

6.1. The AMC model

Figure 4 illustrates a simplified, prototypical model and its dependencies and stereotypes derived from the meta-model. The following section describes each of the elements that make up the AMC model:

- **AMC <Stereotype=Module>** A module which contains AMC services and application functionality.
- **Policy Based Management System <Stereotype=Core>** This is the nucleus of the AMC and defines the behaviour for the AMC by handling the policy system of the AMC. This is contained within the AMC module.
- **Fault Management <Stereotype=Application>** The overall application across an AMC overlay network that handles fault management. Application modules running on each AMC in an overlay make up one application. The application is located outside the AMC module.
- **FM Init, Fault Formatting, Fault Analysis, Fault Detector, Fault Reporting <Stereotype=Application Module>** The Fault Management (FM) application modules (as listed above) handle the AMC's role in the overall fault

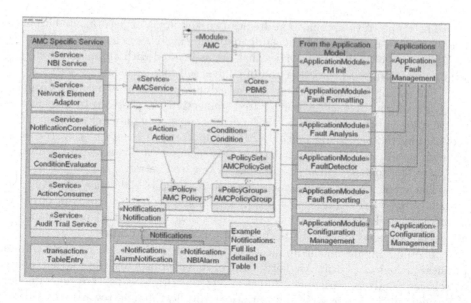

FIGURE 4. A simplified AMC model

management activities of the overlay network. Application modules are down-loaded as they are required by means of the Code Distribution Service. In case that an Application Module is updated it should also be replaced in the AMC (at least at design time). They are described in more detail in Section 6.2. These are contained within the AMC module.

- **Configuration Management <Stereotype=Application>** The overall application across an AMC overlay network that handles configuration management. Application modules running on each AMC in an overlay make up one application. The application is located outside the AMC module.
- **Configuration Management <Stereotype=Application Module>** The Configuration Management (CM) application module co-ordinates the AMCs role in the overall organisation of the overlay network and provides information about the topology of the Madeira management system. Application modules are downloaded as they are required by means of the Code Distribution Service. In case that an Application Module is updated it should also be replaced in the AMC (at least at design time). This is contained within the AMC module.
- **(AMC)Policy <Stereotype=Policy>** The AMC supports a number of policies which govern the behaviour of the management of the NE within the system. One example, used in the Madeira system, is a policy that tells that the node how to react when an Alarm notification is received and the measures that need to be taken given the particular circumstance of the AMC. It may be necessary to re-format the alarm based on a set of rules defined in the policy

(see below) or it may be necessary for the alarm to correlate with a previously received alarm.

- **(AMC)Condition** <**Stereotype=Condition**> An example of a condition supported in the AMC would be the receipt of an "Alarm notification" reporting the loss of the connectivity of the node. This is contained within the AMC module.
- **(AMC)Condition Evaluator** <**Stereotype=Service**> An example of a condition supported in the AMC would be: it can be waiting for receipt of an "Alarm notification" reporting the loss of the connectivity of the node. This is contained within the AMC module.
- **(AMC)Action** <**Stereotype=Action**> The action would refer to an action identifier of the action that needed to be enforced such as: "formatAlarm". The action can also specify parameters that need to be used for during the enforcement of the action. This is contained within the AMC module.
- **(AMC)Action Consumer** <**Stereotype=Service**> An example of a condition supported in the AMC might be the action consumer would cause the "Fault Formatting" application module to be executed (see Section 6.2). This is contained within the AMC module.
- **(AMC)Policy Set** <**Stereotype=Policy Set**> The policies related to Fault Management are grouped in a Set called 'Fault Management Policies'.
- **(AMC)Policy Group** <**Stereotype=Policy Group**> There can be different kinds of Fault Management policies depending on the type of the fault that we want to manage. Each type must be defined as a different group that can have different processing strategies depending on its main focus. The processing strategy refers to the order in which the policies have to be activated.
- **(AMC) Notification** <**Stereotype=Notification**> Notifications allow the AMC to communicate with other AMCs, its platform and external entities. The following table lists some of the notifications that are used by the AMC.

TABLE 1. Selection of AMC Notifications

Notification	Description
AlarmNotification	Madeira alarm format used by the FM Application.
NBIAlarm	Used to report an alarm northbound towards the OSS.
NetworkElementIndication	A change has occurred on the network element.
NetworkElementAlarm	An alarm has been raised on the network element.
NEConfigurationEvent	Changes have occurred in the configuration.

- **Notification Correlation Service** <**Stereotype=Service**> The Notification Correlation service provides generic functions for correlation of notifications. It is heavily used by the Fault Management application.
- **Audit Trail Service** <**Stereotype=Service**> The Audit Trail service records the notifications received and actions taken as a result of the notifications.

- **Transaction** <**Stereotype=Transaction**> A transaction is the record of the information that is stored by the Audit Trail service.
- **Northbound Interface Service** <**Stereotype=Service**> This service acts as the AMC's mean of communication with an external OSS. Its functions include:
  - Receipt of notifications from the CM and FM application and forwarding to the external OSS.
  - Receipt of commands from an external OSS and forwarding them to the appropriate area of the AMC.
  - Provision of information to the external OSS about the network topology which it receives from the CM Application.
  - Handling the introduction of new policies from the external OSS into the AMC.

6.2. The Fault Management Application Model

We took our task one step further and looked more closely at the use of the meta-models for the overall mesh network management applications, in particular its use in the Fault Management (FM) area of our scenario. The Madeira Fault Management application uses a completely decentralised approach: every AMC gathers as much information as possible before forwarding an alarm to a fellow AMC in the network (as opposed to standard centralised approaches, where all correlation and analysis tasks are centralised in one single or a few entities). This greatly removes burden from the higher-level entities in the network [28].

Structurally our FM application splits into five central parts, which have been modelled as AMC application modules (see fig. 5):

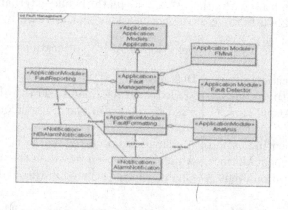

FIGURE 5. Structural Model of Fault Management

- **FMInit** <**Stereotype=ApplicationModule**> The FMInit module is used to realise complex setup tasks for the FM application. Since no such setup tasks are necessary in our concrete scenario this application module has not been mapped to code in our initial prototype application.

- **FaultDetector** <Stereotype=ApplicationModule> If monitoring of some sources of faults involves complex tasks (like permanent polling) a FaultDetector module can be used to accomplish this. Like the FMInit module the FaultDetector module did not prove necessary for the scenario prototype.

- **FaultFormatting** <Stereotype=ApplicationModule> The Madeira FM application handles problems from various sources on different levels of the Madeira platform, but it cannot rely on these notifications to be in a certain format. The FaultFormatting module therefore reformats these source alarms ("raw alarms") into a common Madeira alarm format. More details on the FaultFormatting application can be found below.

- **Analysis** <Stereotype=ApplicationModule> The Analysis module analyses incoming alarms and will perform alarm correlation (using the AMC Notification Correlation service). The Analysis module can be seen as the heart of the FM application. Ref. [28] provides a detailed description of the Fault Analysis and Fault Correlation system.

- **FaultReporting** <Stereotype=ApplicationModule> When alarms are fully analysed on a single AMC (i.e. when a full consolidated view of a problem is established) the AMC forwards the alarm either to one of its fellow AMCs in the network, or to the OSS using the NBI. To decide on that is one of the main responsibilities of the FaultReporting module.

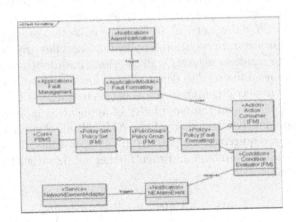

FIGURE 6. Structural Model of Fault formatting

Figure 6 shows one of the aforementioned application modules, the FaultFormatting module, in more detail: the FaultFormatting is centred around special policies (formatting policies), which contain the concrete rule set necessary to reformat a certain low-level alarm into Madeira alarm format. These formatting policies are interpreted by specific FM Condition Evaluators and Action Consumers. The actual FaultFormatting module is invoked by the Action Consumer, it takes a certain rule set in predefined format and a source alarm, and produces a Madeira alarm from it. This alarm is then forwarded to the Analysis module for further processing.

Since all specific formatting rule sets are given to the FaultFormatting module by means of policies the application module itself is held completely generic. Therefore new formatting rule sets for new types of hardware can be introduced during runtime.

6.3. Model Development

The AMC and FM application models were used as a basis for initial prototypical implementation of the AMC and the network management applications. This implementation was initialised under the MDA methodologies but due to its constraints, outlined in [25], and given the project timelines it proved necessary to move to a more conventional software development process. Given this set of circumstances, only a subset of the AMC and application development could use the models presented above. The models were used to generate code stubs and the resulting developments were then hand-coded hence the models provided a useful structuring tool and "map" during the more detailed implementation of the functionality.

The resulting developed AMC and application addressed the key mesh network management requirements discussed in Chapter 4. Distributed control is addressed by the presence of application modules in distributed AMCs which come together in an AMC overlay network to form the collection of application modules known as the application. In addition, application modules are modular in nature and do not need to reside in all AMCs all the time. In order to cope with this flexibility policies are used in order to define some criteria to download and distribute these modules. Symmetric communication between AMCs is handled by the use of notifications. Policies and the modularity of application modules allows for self organisation and adaptive behaviour of the AMC. Table 2 summarises the means by which the main requirements identified in Chapter 4 have been represented as part of the Madeira meta-model and then further validated within the AMC and application models.

TABLE 2. Addressing Requirements

Requirement	Element
Distributed Control	Application Module, Policy, Application
Symmetric Communication	Notifications
Self-organisation	Policy, application modules.

7. Conclusions

The Madeira project addresses the network management of next generation networks in particular the Peer-to-Peer paradigm. The challenges in managing such networks include: distributed nature of network elements, heterogeneity, self-organisation and communication. While the solution has been specifically applied to P2P networks, by addressing the characteristics above, we can expand its relevance towards emerging autonomic networks.

In this paper, we sought to capture the key concepts and semantics that apply to the network management of a P2P network and its related challenges in a meta-model. The very nature of the meta-model allows the problem to be extracted into a semantic form from which it can be visualised without being encumbered by the minute details that need to be solved. Current meta-models that are used to describe the management of networks address the more centralised network management architectures and fail to address the characteristics that identify a mesh network, which include distributed behaviour.

By deriving our meta-model from the meta-meta model, MoP, behaviour was introduced via policies and events from the very top layer. The MAMA meta-model was then extended to meet the characteristics introduced by MoP. The required characteristics of self-management, adaptive behaviour and communication are successfully included in the Madeira meta-model via the use of policies, notifications and applications. The distributed behaviour of an application running as an overlay across AMCs across a mesh network is captured using the concepts of application, application module and also the policy's ability to enable functional flexibility.

The application of the meta-model is verified in our instantiation of the AMC and related FM application model. It has been demonstrated that the semantics and relationships in the meta-model can be used to model a managing entity of a P2P network and its application. Our Fault Management application uses the AMC model and a policy-based approach to master a challenging scenario from the area of wireless networking, which can be seen as a metaphor for many other NGN scenarios. The study of [25] discusses the overall Model Driven Approach that was taken in Madeira and details the barriers that have impeded the transformation from model to automatic and full AMC and application software instantiation.

Future work would include the evaluation of emerging MDA or other appropriate technologies that might be used to implement a complete solution for the AMC and hence reinforce the validity of the meta-model. The meta-model's applicability to the more complex networks such as Autonomic Networks would also be considered.

References

[1] E. Pitoura, G. Samaras, G. Samaras, *Data Management for Mobile Computing.* (1997).

[2] J. Strassner, J. Kephart, *Autonomic Systems and Networks: Theory and Practice.* NOMS Tutorial, 2006.

[3] Y.Yemini, S. daSilva, *Towards programmable networks.* IEEE International Workshop on Distributed Systems: Operations and Management (DSOM 1996).

[4] AG. Ganek, TA. Corbi, *The dawning of the autonomic computing era.* IBM Systems Journal, **42:1**, 2003.

[5] K. Lyytinen, G. Rose, R. Welke, *The Brave New World of development in the internetwork computing architecture (InterNCA): or how distributed computing platforms will change systems development.* Information Systems Journal, **8:3**, 241-253.

[6] J. Risson, T. Moors, *Survey of Research towards Robust Peer-to-Peer Networks: Search Methods.* Technical report UNSW-EE-P2P-1-1, University of New South Wales, Sydney, Australia, 2004.

[7] M. Zach, D. Parker, J. Nielsen, C. Fahy, R. Carroll, E. Lehtihet, N. Georgalas, R. Marın, J. Serrat, *Towards a framework for network management applications based on peer-to-peer paradigms.* NOMS 2006.

[8] M. Roussopoulos, M. Baker, et al., *2 P2P or Not 2P2P?* The 3'rd International Workshop on Peer to Peer systems, San Diego, USA Feb 26-27, 2004.

[9] C. Shirky, *What is P2P ... and what isn't.* http://www.openp2p.com/pub/a/p2p/2000/11/24/shirky1-whatisp2p.html.

[10] European Celtic initiative. http://www.celtic-initiative.org/.

[11] Madeira project. http://www.celtic-madeira.org.

[12] IETF, *Configuring Networks and Devices with Simple Network Management Protocol (SNMP).* Network Working Group, 2003 RFC 3512.

[13] ITU-T, *Common Management Information Protocol Specification.* ITU-T Recommendations X.711, 1997.

[14] DMTF, *Desktop Management Interface Specification.* DSP0005, DMTF, Version 2.0.1s, January 10, 2003.

[15] DMTF, *WBEM Discovery Using Service Location Protocol.* DSP0205, DMTF, Version 1.0.0, January 27,2004.

[16] DMTF, *WBEM URI Mapping Specification.* DSP0207, DMTF, Version 1.0.0l, January 25, 2006.

[17] OMG, *Meta Object Facility(MOF) Core Specification.* Version 2.0, OMG, January 2006.

[18] TMF, *NGOSS Architecture Technology Neutral Specification Metamodel.* Annex TMF053D, Version 1.0, February 2003.

[19] N. Georgalas, *The Model of Object Primitives (MOP).* Succeeding With Object Databases : A Practical Look At Today's Implementation With Java and Xml, October 2000, 464 pages, ISBN 0-471-38384-8.

[20] S. van der Meer, *Middleware and Application Management Architecture.* PhD Thesis, Berlin, Germany, September 25, 2002.

[21] K. Raymond, *Reference Model of Open Distributed Processing (RM-ODP): Introduction.* Proc. of the International Conference on Open Distributed Processing, ICODP'95, Brisbane, Australia, 20 - 24 February, 1995.

[22] TINA-C, *Computational Modeling Concepts*. TINAC Deliverable, TINA 1.0, Version 3.2, Archiving Label TP_HC.012_3.2_96, TINAC, May 17, 1996.

[23] OMG, *The Common Object Request Broker: Core Specification*. Version 3.0.3, OMG, March 2004.

[24] DMTF, *Common Information Model (CIM) Specification*. DMTF, Version 2.12.0, April, 2006.

[25] OMG, *MDA Guide*. DMTF, Version 2.12.0, April, 2006.

[26] R. Carroll, C. Fahy, E. Lehtihet, S. van der Meer, N. Georgalas, D. Cleary, *Applying the P2P paradigm to mamagement of large-scale distributed networks using a Model Driven Approach*. NOMS 2006.

[27] T. Clark, A. Evans, P. Sammut, J. Williams, *Language Driven Development and MDA*. Xactium Limited, http://www.bptrends.com/publicationfiles/10%2D04%20COL%20MDA%20%2D%20Frankel%20%2D%20Xactium%2Epdf.

[28] ITU-T Recommendation X.733 *Information Technology Open Systems Interconnection Systems Management: Alarm Reporting Function*. ITU-T.

[29] Leitner M., Leitner Ph. ,Zach M., Fahy, C., Collins S *Fault Management based on peer-to-peer paradigms: A case study report from the CELTIC project Madeira*. Integrated Management 2007, Munich, April 2007.

C. Fahy, M. Ponce de Leon and S. van der Meer
Waterford Institute of Technology
Waterford
Ireland
e-mail: cfahy@tssg.org

R. Marin, J. Vivero and J. Serrat
Network Management Group
Universitat Politcnica de Catalunya
Barcelona
Spain
e-mail: rmarin@nmg.upc.edu

N. Georgalas
BT Group
Ipswich
UK
e-mail: nektarios.georgalas@bt.com

P. Leitner
PSE - Program and System Engineering
Siemens AG Austria
Vienna
Austria
e-mail: philipp.leitner@siemens.com

S.Collins
Ericsson R & D Ireland
Athlone
Ireland
e-mail: sandra.collins@ericsson.com

B. Baesjou
Telefónica I & D
Spain
e-mail: baesjou@tid.es

Whitestein Series in Software Agent Technologies, 63–80
© 2007 Birkhäuser Verlag Basel/Switzerland

Autonomic Communication with RASCAL Hybrid Connectivity Management

Dominic Greenwood and Roberto Ghizzioli

Abstract. This paper presents an approach to manipulating available hybrid connectivity to autonomically maximise the potential for sustained connectivity in the event of path disruptions. The approach is documented in terms of the features, architecture and deployment modes of an autonomic communications module, termed RASCAL. This module employs software-agent logic supported by a state-of-the-art policy engine to dynamically determine best options for packet transmission over available infrastructure and ad-hoc connections.

Keywords. autonomic, hybrid, ad-hoc, contingency, policy, agent.

1. Introduction

The ability to seamlessly communicate when mobile is now, for many, an inescapable component of day-to-day life. It is of course the electronic communications revolution which has brought about this reality; one where in many respects we simply cannot perform many common tasks without access to communicative devices including cellphones, PDAs, laptops and GPS. This fact is especially resonant in environments where communication is critical to sustaining coordination between individuals that need to remain *always-best-connected* anywhere, anytime, using any available network technology and with the maximum quality and capacity on offer. We consider key examples of such environments to include those where human life is a critical concern, such as sites of natural disasters (e.g., tsunami, hurricanes, earthquakes, floods, forest fires, etc.), major incidents [8, 13] (e.g., plane crashes, multi-vehicle road traffic accidents, building fires, etc.) and theaters of military operations. In all of these there is very often a pressing need to communicate information between individuals, whether they be in either localised, or widely distributed groups. A straightforward example is the coordination of ad-hoc teams of rescue workers that need to share multiple forms of information (i.e., audio, video, sensor data, medical data, etc.) while operating in the field.

With this work we thus aim to address the problem of maximising the assurance that communication will remain established even when communicative channels are disrupted due to environmental events. Specifically, we propose one contribution to solving this problem: an autonomic communication system providing real-time, secure, bidirectional communication of data messages from source to destination(s) while remaining agnostic to the devices, networks or carriers required to transfer the information. The reported technology prototype draws on concepts defined by many researchers and practitioners in the field of autonomic communication [1, 12].

The prototype is termed RASCAL[1] (*Resilience and Adaptivity System for Connectivity over Ad-hoc Links*). RASCAL is a novel middleware communication mechanism that automatically ensures (to such degrees as are possible within the operating environment) the continued operation of ubiquitous application services where communication may be subject to disruption. This requires some migration of message handling intelligence into user devices to allow iterative delivery decision-making throughout the communication route. RASCAL thus shifts the burden of tasks such as configuration, maintenance and fault management from users to a specialised self-regulating subsystem. A local policy engine is used by each RASCAL deployment for *self-configuration* purposes, allowing autonomic adjustment of behaviour in accordance with environmental changes. RASCAL is also *self-optimising* as it monitors network resources and adapts its behaviour to meet the end-user and application service needs, i.e., automatically handing-over sustained sessions between WLAN and Cellular connections to maximise the always-best-connected goal. Furthermore, RASCAL is also *self-healing* when managing multiple bearer technologies, such as WLAN, 3G or Bluetooth; for example, switching to an ad-hoc connection in the temporary absence of an infrastructure connection. Finally, RASCAL also offers an intuitive interface allowing the user to inspect ongoing activities, decisions and internal state of the system.

The remainder of this paper is organised as follows: Section 2 discusses some relevant previous work on autonomic communication. Section 3 then presents the autonomic features of RASCAL and Section 4 presents the RASCAL architecture. Section 5 presents some initial results from laboratory-based experimentation and Section 6 illustrates a real-life scenario within which RASCAL has been evaluated. The paper is concluded in Section 7.

2. Related Work

As mentioned in the previous section, there are a variety of published studies available regarding the deployment of autonomic communication and network management systems in disruptive environments. The majority of these tend to consider quite specific aspects of the domain. For example, when addressing the networking

---

[1]RASCAL has been designed and implemented as a deliverable of the European Union 6th Framework Program Palpable Computing project (PalCom) - IST-002057.

aspect most papers focus on either infrastructure or on ad-hoc networks (MANET) without considering the coordinated use of both concurrently. A specific case in point is the work of Chadha *et al.* [20] who present an autonomic system developed under the U.S. Army CERDEC DRAMA (Dynamic Re-Addressing and Management for the Army) program deployed in military scenarios. In particular, this system only addresses mobile ad-hoc networks without consideration of potentially available infrastructure networks.

Those papers that deal with hybrid networks consisting of a combined infrastructure and mobile ad-hoc connectivity, few then consider either the requirements and influence of the user applications running over the autonomic communications subsystem, or the roles of end-users in deployment scenarios.

A good example of hybrid network management is provided in Hauge *et al.* [21] who present an interesting approach to the combined use of 3G cellular and ad-hoc networks. They conclude that hybrid networks provide the opportunity to transmit service data to a higher percentage of interested mobile terminals than when using only an infrastructure network. Nevertheless, the role of their user-level service (multicast) within hybrid networks is not considered. The same can be said of the work of the *Delay Tolerant Networking Research Group* [3] chartered as part of the Internet Research Task Force (IRTF). This group primarily focuses on network aspects providing end-to-end connectivity in disruptive environments without considering how application contexts influence the achievement of connection/delivery goals. We address this issue as a component of the RASCAL usage-aware approach (see section 3).

Another important work in this area is that of Kappler *et al.* [22] who use a policy-engine to address hybrid network composition. Although this aspect is in line with our approach, the authors do not consider the discovery of relevant network nodes, and policies are only used for the composition of network devices, whereas our approach also considers the possibility of user-level service composition.

The notion of Unified Messaging (UM), as reported in van der Meer *et al.* [24] for example, is also closely related to our work from the perspective of supporting both fixed and mobile users with universal access to communication services. The central concept of UM is the capability of the messaging system to select the most appropriate terminal or application for an incoming message according to availability, status and other parameters. A UM system is designed to adapt terminals to different kinds of services via content adaptation processes guided by user rule policies. However, the particular approach documented in [24] identifies CORBA as a suitable means of engineering the middleware software for handling UM; an approach, in our opinion, not entirely suitable for application in pervasive environments where seamless mobility is a paramount issue and small footprint devices are the norm, as is particularly the case in disruptive environments.

Additionally, published work relating to the use of ambient and pervasive technologies in disruptive environments or disaster management tends to focus more on service composition or other application-oriented aspects without considering the underlying networks issues. Such an example is the reported work of Kristensen *et al.* [13] which focuses on IT support in major incidents, such as the use of bio-monitors, patient identification and collaboration tools for response units, without considering how these applications could, or should, behave in the presence of network disruptions. The RASCAL system reduces this gap between specific autonomic aspects purely based on the network management and the autonomic aspects based on the user-level services deployed in disruptive environments.

3. The Autonomic Features of RASCAL

The RASCAL software system is a middleware communication layer offering vertical interfaces to communicative applications (typically user-driven) and to low-level network bearer modules. The purpose of the layer is to intercept all, or a selection of, application-specific messages passing through the local communication subsystem of a device in accordance with a set of dynamically configurable policies defining the prioritised actions to take regarding forward routing of the messages. These policies are thus used to guide the autonomic decisions that the RASCAL autonomic controller can take in response to events sensed from the environment. The most straightforward example is the detected failure of an infrastructure connection (say WLAN), whereby a predefined obligation policy may mandate that RASCAL re-route messages via an alternative network technology (say Bluetooth) to either their final destination or another node with an active infrastructure connection. An example of such a policy is described in Section 6.

This technology thus improves the potential for communicative applications to remain connected when deployed in environments subject to disruptive behaviour. When RASCAL is deployed within the local communications stack of a device, we term the device as having become *RASCALised*.

The primary features of RASCAL fall into two classes: *connection-aware* and *usage-aware* communication.

Connection-aware communication implies the ability of RASCAL to be aware of all available (active and inactive) network connections, their parameterisation and performance characteristics (e.g., Quality of Service (QoS) characteristics). This set of autonomic behaviours is triggered by changes in network resources, with some of the most significant operations being:

- **Network handover:** The capability of dynamically switching from one network type to another when communicating with other devices. This decision can be taken based on the reachability of a device over different infrastructure or ad-hoc networks and on network availabilities. For example, we can consider handovers from an infrastructure technology (e.g., UDP) to an ad-hoc technology (e.g., Bluetooth) when communicating with a device in the local neighbourhood.

- **Routing optimisation**: The capability of enhancing multi-hop routing among network nodes. Parameters which affect these decisions can be based on several QoS parameters such as response delay, nominal and available bandwidth between network nodes, transmission errors, etc. For example, a self-optimisation feature of RASCAL fitting into this group is the proactive evaluation of the transmission delay between two interoperating devices and consequently the use of an alternative path to reach the same target node.

Other behaviours within this category include those acting in response to fail-overs or high network load, etc.

As mentioned, RASCAL also offers *usage-aware* communication consisting of a set of autonomic behaviours related to the usage of deployed user-level services. For example, a group of rescue workers are on the site of a major accident with human casualties and must constantly maintain communication both with one another and with a response unit concerning their findings and the positions of injured people. Due to the potentially disruptive nature of the environment network availability may be intermittent, but the goal to reliably deliver communication must persist. In this scenario the typical autonomic decisions to be taken include:

- **Transmission contingency**: The capability of providing alternatives to the default means of transmitting a message. This specifically includes making the best possible use of multi-technology transmission paths including cellular networks, IP infrastructure networks, satellite systems, MANETs, etc. An important parameter in these decisions is the importance of the message to be sent. An example is simultaneously sending high priority messages via two or more different technologies, and therefore routes, to improve the chances that they are successfully delivered to target nodes. The use of extra resources is justified by the importance of the content to deliver.
- **Content adaptation**: Adapting the content of a message (or stream). These decisions can be based on several parameters such as the number and the importance of the messages/streams to be sent. Examples include applying a codec to reduce the bandwidth consumed by a video stream, or simply stripping out the audio component and sending this in lieu of the video.
- **Deferred service provisioning**: Waiting until a connection is available before making routing decisions to mitigate uncertainty relating to the choice of optimal technologies or paths. Also in this case the parameters able to trigger this type of decisions are the volume and the significance of the information to be sent. This includes the need to buffer messages while awaiting a connection.
- **Role management**: Specifying user defined conditions which must be met before taking a particular action. A parameter which affects these decisions is the role of the end-user in a particular scenario. For example within a major incident response workers are divided into response units, structured in a hierarchical manner. In this scenario we can envisage policies which ensure a message (e.g., notification of an event) is sent to the right recipients in the role hierarchy (e.g., escalating to the fire officer).

All of these capabilities are controlled by user-definable policies that may be simply and rapidly specified/modified on-the-fly by either the user or local/remote automated routines.

Additionally, one other important feature of RASCAL is the notification, inspection and control interface available to the end user. This GUI specifically allows the user to remain aware of decisions made by the autonomic controller and to affect them if necessary. More details on this are provided in Section 4.4.

4. The Architecture of RASCAL

The RASCAL software architecture is depicted in Figure 1 and is specifically designed to be compliant with the standard autonomic control loop [2], in that it:

- *collects* sensory information from the system and the external world,
- *decides actions* using a policy engine
- and *effects* those decisions to affect system behaviour.

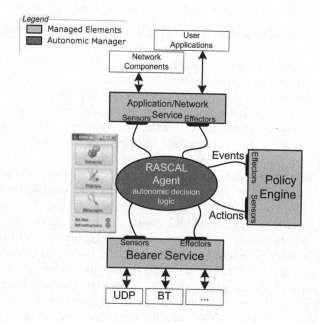

FIGURE 1. The RASCAL software architecture.

The architecture consists of three managed elements, which are in fact software services: an applications service, a networks service and policy service, all of which will be described later. Also present is the core decision control logic in the shape of an autonomic manager (RASCAL software agent), a policy engine and the user interface control.

The RASCAL software is designed to interact with different layers of the user-device communication stack. In the default case it interacts with lower-layer communications components dealing with bearer technologies, e.g., UDP, Bluetooth, etc., and with upper-layer components dealing with, for example, routing protocols, discovery and services/applications. Other components may also interface to RASCAL through provided software interfaces.

As we believe a highly beneficial means of implementing the control logic of an autonomic manager is with a software agent, we elected to design the RASCAL architecture for implementation as a JADE (Java Agent Development Framework) software system. JADE ^{2} is an open source software agent platform and development environment [4]. Thus the RASCAL autonomic manager is designed for deployment as a JADE agent and the managed element interfaces are deployed as JADE kernel services, both of which are executable within the JADE runtime. In general terms a *software agent* is defined by Wooldridge [15] as a *"computer system that is situated in some environment and that is capable of autonomous action in this environment in order to meet its design requirements"*. A JADE kernel service is defined by Bellifemine *et al.* [4] as a *"software component which implements platform level features that can be grouped together according to their conceptual cohesion"*.

4.1. The Managed Elements

The RASCAL agent interacts with the external world via JADE kernel services. Each service is controlled through sensors and effectors. Effectors produce actions relating to instructions received from the RASCAL agent; they implement the *Command* design pattern [14]. Sensors collect information from the external world and provide it to the RASCAL agent for processing; they implement a simplified version of the *Half-Sync/Half-Async* design pattern [5].

The RASCAL system contains three JADE kernel services:

The Network/Application Service. This is used by the RASCAL agent to interact with upper-layer network services, such as routing and discovery, and application services facing the user. Messages passing through this interface may be inspected by RASCAL to determine whether any action is necessary by the autonomic manager in accordance with specified policies. A selection of interfaces is available allowing communication with a broad range of network services and applications.

The Bearer Service. This is used by the RASCAL agent to interact with the lower-layer bearer interfaces to both infrastructure and ad-hoc network endpoints. Messages passing through this interface may be inspected by RASCAL to determine whether any action is necessary by the autonomic manager in accordance with specified policies. A selection of interfaces is available for a broad range of network technologies.

---

^{2} JADE documentation and software is available from http://jade.tilab.com/

The Policy Service. This is used by the RASCAL agent to interact with the local policy engine (see Section 4.3). In brief, this engine possesses the operational rules that must be applied to control (or not control) the way in which messages are treated by RASCAL. The RASCAL agent uses these policy rules to effect this control.

4.2. The Autonomic Manager

This is the RASCAL software agent that controls the RASCAL system such that it exhibits the following autonomic behaviours:

Sensing. By installing sensors in the managed elements the agent is able to monitor new application messages to be sent, new messages received from the network or new network status events. Data is collected both asynchronously (the managed elements notify of a status changing) or synchronously (the agent explicitly requests for information).

Compiling Knowledge. Intercepted messages and received events are used to create an internalised model of the external world. This compiled knowledge base contains information such as statistical flow data, historical fault logs, active and treated faults, discovered devices and the services those devices provide.

Decision Control. Whenever the internal knowledge base is updated, the RASCAL agent triggers a call to the local policy engine which dictates the policy constraints that must guide decision-making by the agent. This aspect is discussed in more detail in Section 4.3.

Proactivity. Actions can be executed by the RASCAL agent immediately, or postponed until some point in the future. In order to schedule such future actions RASCAL implements a model of time allowing proactive planning.

4.3. Decision Making

A core aspect of the RASCAL system is its reasoning system. The RASCAL agent receives messages from user-level applications and probes the environment using the previously discussed managed element kernel services. Decisions on how to treat the received messages are made locally using *policies*; a set of constraint rules governing system behaviour. One of the goals of RASCAL is to provide the end user with an easy means of authoring policies that will control the various autonomic features (see Figure 2). In order to modify the RASCAL behaviour at runtime, these policies are dynamically loaded when the device is running.

Policies are not coded directly within the agent behaviours. To be more flexible, the agent uses the Policy Service managed element, shown in Figure 1, to issue events to a policy engine and wait for a set of recommended actions to perform. The particular policy server employed by RASCAL is Ponder2[3] [16, 17], used as a local library, which uses an XML-based policy description language to define events and policies to be processed by the Ponder2 policy engine. The result of a policy is an action the RASCAL agent has to perform. Currently, RASCAL deals with *obligation policies*. An obligation policy is an Event Condition Action (ECA)

---

[3]see http://ponder2.net/

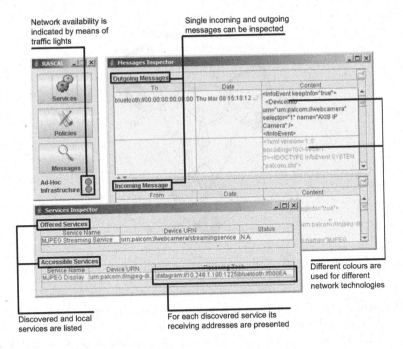

FIGURE 2. The RASCAL GUI.

rule in the deontic sense [18]. Given E and C is true, it is obligatory that the agent performs A.

4.4. Graphical User Interface

The RASCAL user interface is designed for control and inspection using event-based interaction with the RASCAL agent. The main panel, shown in Figure 2, indicates the status of the various network interfaces available on the local device and a set of buttons to open inspection views. One of these views provides a list of all remote devices interacting with user-level services running on the local device. Additionally, a second view is dedicated to message inspection and a third to inspecting, editing and controlling policies definitions. Currently, the RASCAL GUI has been implemented to work only on laptops or personal computers but it can be easily adapted to other consumer devices like PDA, smartphones, etc.

5. Laboratory Experimentation

This section presents preliminary laboratory experiments conducted in order to validate RASCAL capabilities in terms of ensuring connectivity remains established even when some channels are disrupted.

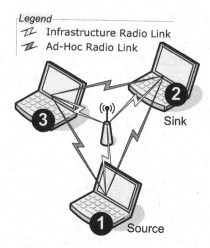

FIGURE 3. Laboratory experimentation setup.

The experimental setup for this evaluation was composed of three laptop nodes, each equipped with WLAN and Bluetooth adapters. Each node could communicate with the others using either of the two available network technologies (see Figure 3).

Each node was also equipped with the PalCom communication stack, which provided discovery services and multi-hop routing capabilities via the DSDV [23] routing algorithm. Selected test applications were deployed on top of the PalCom communication stack: The application running on node-1 (the source) was to send heartbeat messages every 3 seconds to node-2 (the sink), once it had been discovered. The application running on node-2 was capable of receiving and counting incoming messages. In this scenario, node-1 could reach node-2 directly, or via node-3. It could also occur that messages could be transferred via hops across different bearers.

To add uncertainty to the experiments, aperiodic network failures were simulated. Each node was equipped with a failure generator which, based on a mathematical model, blocked the transmission of messages over a particular network adapter for a certain time. The mathematical model was based on the *Markovian property* that the probability of the occurrence of an event does not depend on the history of previous events. Based on this property, technology failures were simulated with an occurrence rate equal to the inverse of the λ parameter of a *negative exponential* distribution. Furthermore, the duration of the failure was simulated using the *Erlang-k* distribution. The expected average and standard deviation failure duration time were used to define the distribution. Table 1 shows the average failure occurrence time and the relative duration for each bearer endpoint. The choice of these values was based on experience gained when performing real-world evaluations (see Section 6).

TABLE 1. Table of average stochastic failure occurrence time and duration per network bearer technology

Technology	Rate (mins)	Duration (mins)
UDP	2.5	2 ± 1
Bluetooth	1.25	2 ± 1

In the experiment, every node was also equipped with a software component named *Failure Generator* which simulated the unavailability of a particular network bearer. In particular, using the parameters presented in Table 1 the component generated failure events that described when and for how long a network adapter had to be considered deactivated. When the event occurred, the failure was simulated and the node prevented from sending messages using that particular adapter. Each time an event was consumed, a new one was immediately generated. The failure generator reactivated an adapter when the failure duration time elapsed.

The entire experiment consisted of 20 runs of 10 minutes each. Every experiment contained a stochastic number of failure events.

To measure the capabilities of RASCAL to ensure that communication remained established even when channels were disrupted, experiments were conducted both with and without the RASCAL component deployed. A node without RASCAL was only capable of communicating using a UDP bearer, with no ability to autonomically adapt. The measured output variable was the *number of messages successfully delivered to the sink* (node-2).

5.1. Results

Figure 4 shows the value of the measured output variable over 10 minutes when analysing a single experimental instance. As can be observed, after the 10 minute cycle the number of messages delivered with RASCAL enabled was substantially higher than without it. In fact, during this period of time several network failures occurred, but nodes with RASCAL deployed continued to discover one another with heartbeat messages sent continuously using different routes. On the other hand, with nodes without RASCAL deployed when a failure occurred the sink was no longer discovered and no heartbeat messages were therefore transmitted from the source to the sink (this is represented by the flat parts of the dotted line in Figure 4).

A more significant result is given by Figure 5 which shows the value of the output variable over all 20 considered instances. This box-plot shows that the average number of messages successfully delivered to the sink when RASCAL was deployed, were definitely better than without it (339 messages against 267). Naturally, for simple instances with no failures the presence of RASCAL was irrelevant. In fact in both cases the maximum number of delivered messages is almost the same. The results change however when many failures occurred since the minimum of this output variable over 20 instances without RASCAL is 110 and with

D. Greenwood and R. Ghizzioli

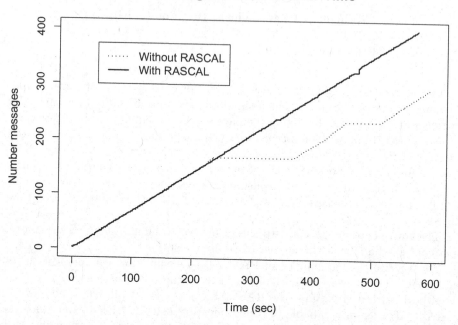

FIGURE 4. Number of messages successfully delivered to the sink
in an experimental instance.

RASCAL is 221, i.e., system performance improved by more than 100%. This allows us to conclude that the more disruption occurs to the network infrastructure, the more benefit is provided by the presence of RASCAL technology - thanks to its capability of utilising alternative paths over different technologies to deliver messages.

To be certain of the real significance of the obtained results over 20 instances the Wilcoxon Paired Rank Sum Test was applied. This test stated with a confidence level higher than 95% (p-value = 0.0008909) that the improvements generated by RASCAL are statistically significant.

6. Real World Evaluation

RASCAL has recently been integrated into the iterative, participatory design process practiced in the PalCom project[1]. We are currently carrying out experiments with end users in major incident emergency response scenarios. This section

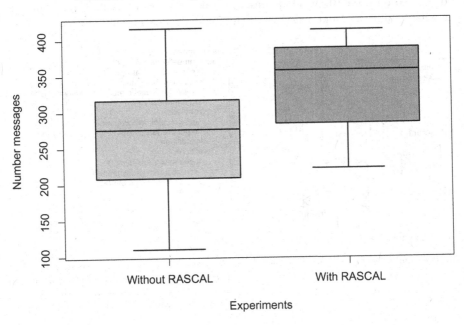

Number of messages successfully delivered to the sink

FIGURE 5. This box-plot represents the number of messages successfully delivered over the 20 experimental instances, with and without RASCAL.

presents an overview of one of the mocked-up situations of a real world major incident conducted recently.

Within this experiment a building fire is considered. This particular scenario demands fast and decisive action, often in life-threatening situations. It also requires collaboration between numerous people located in different, often changing areas: personnel at the incident site including firefighters, police and paramedics, at the command centre, in vehicles and others.

Each of the people involved, and many of the vehicles and other equipment, are associated with one or more electronic devices such as radios, biosensors, GPS, health recorders, handhelds, tablet PC, etc. In the fire scenario different devices run different crisis-relevant applications including VOIP clients, instant messengers, map services and other collaborative tools.

The particular scenario is illustrated by Figure 6(a). Three firefighters (FF) are moving in relative proximity to one another attempting to evacuate people from a building on fire. They are using small, wearable *RASCALised* PDAs, each with a built-in camera running a map service. Their duty is to notify the command centre

(CC) and the other local team members of findings related to visited building(s), i.e., the positions of injured people. To do this, they make special marks on the map displayed on their PDA. They may also take pictures to assist the command centre with gaining a visual overview of the incident location and status.

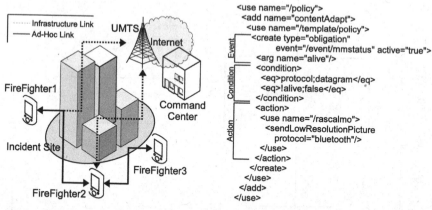

(a) The conducted mocked-up situation.

(b) Example of policy to send low resolution pictures to the command centre towards the ad-hoc network when the infrastructure connection is not working anymore.

FIGURE 6. Real World Scenario.

In terms of connectivity, FF1 and FF2 are connected via both ad-hoc (HOC) and infrastructure (INFR) networks and FF3 only via an ad-hoc connection. In this situation, through the multi-hop capabilities of RASCAL all four actors (the three firefighters and the command centre) are able to communicate with one another. For example FF3 communicates with the command centre via the ad-hoc connection with FF2.

In this particular scenario, the *RASCALised* devices used by the firefighters are equipped with the following policies:

1. IF (infrastructure_connected) THEN send map data to CC and FF via INFR.
2. IF (!infrastructure_connected) THEN send map data to CC and FF via HOC.
3. IF (infrastructure_connected) THEN send high resolution pictures to CC via INFR.
4. IF (!infrastructure_connected) THEN send low resolution pictures to CC via HOC.

Figure 6(b) shows the definition of the final policy in the list presented above. This policy sends low resolution pictures to the command centre when the device is not infrastructure connected. It receives *mmstatus* events which denote whether

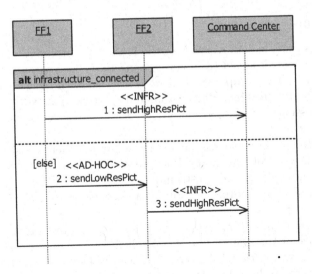

FIGURE 7. Sequence diagram of the actions taken by FF1 to send pictures to the command centre.

a network interface is available or not. This event has two associated parameters: the interface protocol and its current status. The only condition that triggers this policy is that the infrastructure connection is unavailable (see the XML *condition* element). The action *sendLowResolutionPicture* triggered by this policy notifies the RASCAL agent to decrease the resolution of the pictures for the command centre and to send them using the ad-hoc network (see the *protocol* attribute of the *sendLowResolutionPicture* XML element).

In this scenario, FF1 has policies (1) and (3) activated. When FF1 moves into an area where the infrastructure connection fails, this is detected and policies (2) and (4) automatically become active. The RASCAL agent running on the FF1's PDA is thus notified by the policy engine and hands over all communication with FF2 to the available ad-hoc connection. Given the importance of sending images to the command centre and giving the low nominal bandwidth of the ad-hoc channel, pictures are first automatically reduced in quality (i.e., resolution) before transmission.

Later, when FF1 returns to an area with infrastructure network coverage, communications with the command centre are automatically returned to the infrastructure connection with images once again sent in normal, high resolution.

A sequence diagram of the actions taken by FF1 to send pictures to the command centre is shown in Figure 7. The diagram considers the situation both with and without the infrastructure connection.

Further real world experimentation with emergency services is ongoing.

7. Conclusion

This paper has provided an overview of the RASCAL autonomic communications software component designed to manage connectivity in environments subject to disruptive events. The component can be interfaced with device communication stacks and offers features including automated network handover, routing optimisation, transmission contingency, content adaptation, deferred service provisioning and role management.

The key components of each local RASCAL deployment consist of a software agent autonomic control logic and a Ponder2 policy engine with which the user is able to define rules guiding and constraining the agent control logic. The autonomic controller interfaces with three (or more) managed element services, of which the policy engine is one.

The reported laboratory experimentation has demonstrated the operation and performance of the system under controlled conditions, with results proving the intuitive conclusion that RASCAL significantly aids the maintenance of sustained connectivity in disruptive environments. Moreover, the reported real world evaluation offers an insight into how RASCAL has been deployed in an actual setting involving firefighters working collaboratively at an incident site.

As a work in progress, RASCAL remains under a continuous refinement, in particular as feedback from real world evaluations is gathered. Particular ongoing work includes areas that enhance the autonomic capabilities of the system including the incorporation of improvements to the Ponder2 policy language (undertaken as an independent project to RASCAL), and to the autonomic controller logic. The improvements to the policy language include a new means of expression using a form of process-algebra over actions. This allows the expression of a set of actions (i.e., workflow) rather than the current limitation to atomic actions. Further planned improvements relate to the inspectability of the RASCAL system and its use in composing dynamic assemblies of computational devices and services with flexible, self-adaptive communicative connections.

References

[1] D. F. Bantz, C. Bisdikian, D. Challener, J. P. Karidis, S. Mastrianni, A. Mohindra, D. G. Shea, M. Vanover, *Autonomic personal computing*. IBM Syst. J. **42:1**, 2003, 165–176.

[2] IBM, *An architectural blueprint for autonomic computing*. 2006.

[3] S. Farrell, V. Cahill, *Delay and Disruption Tolerant Networking*. Artech House Publishers, 2006.

[4] F. L. Bellifemine, G. Caire, D. Greenwood, *Developing Multi-Agent Systems with JADE*. John Wiley & Sons, 2007.

[5] J. M. Vlissides, J. O. Coplien, N. L. Kerth, *Pattern languages of program design 2*. Addison-Wesley Longman Publishing Co., 1996.

[6] F. Hu, S. Kumar, *The integration of ad hoc sensor and cellular networks for multi-class data transmission.* Elsevier Ad-Hoc Networks Journal, **4:2**, 2006, 254–282.

[7] U. Varshney, S. Sneha, *Patient monitoring using ad hoc wireless networks: reliability and power management.* IEEE Communications Magazine, **44**, 2006, 49–55.

[8] M. Kyng, E. T. Nielsen, M. Kristensen, *Challenges in designing interactive systems for emergency response.* DIS '06: Proceedings of the 6th ACM Conference on Designing Interactive Systems, 2006, 301–310.

[9] Z. Obrenovic, D. Starcevic, E. Jovanov, V. Radivojevic, *An agent based framework for virtual medical devices.* AAMAS '02: Proceedings of the first International joint Conference on Autonomous Agents and Multiagent Systems, 2002, 659–660.

[10] M. Ingstrup, K. M. Hansen, *Palpable assemblies: Dynamic composition for ubiquitous computing.* Proceedings of the seventeenth International Conference on Software and Knowledge Engineering, 2005.

[11] D. Greenwood, M. Calisti, *The Living Systems Connection Agent: Seamless Mobility at Work.* Communication in Distributed Systems (KiVS 07), 2007.

[12] J. Strassner, *Seamless Mobility - A Compelling Blend of Ubiquitous and Autonomic Computing.* Dagstuhl Workshop on Autonomic Networking, 2006.

[13] M. Kristensen, M. Kyng, E. T. Nielsen, *IT support for healthcare professionals acting in major incidents.* 3rd Scandinavian Conference on Health Informatics, 2005, 37–41.

[14] E. Gamma, R. Helm, R. Johnson, J. Vlissides, *Design patterns: elements of reusable object-oriented software.* Addison-Wesley Professional, 1995.

[15] M. Wooldridge, *An introduction to multiagent systems.* John Wiley & Sons, 2002.

[16] N. Damianou, N. Dulay, E. Lupu, M. Sloman, *The Ponder Policy Specification Language.* POLICY '01: Proceedings of the IEEE Workshop on Policies for Distributed Systems and Networks, Bristol, UK, 2001, 18–38.

[17] G. Russello, C. Dong, N. Dulay, *Authorisation and Conflict Resolution for Hierarchical Domains.* POLICY '07: Proceedings of IEEE Workshop on Policies for Distributed Systems and Networks, Bologna, Italy, 2007.

[18] R. J. Wieringa, J.-J. Ch. Meyer, *Applications of deontic logic in computer science: a concise overview.* Deontic logic in computer science: normative system specification, John Wiley and Sons Ltd., 1993, 17–40.

[19] D. Siegrist, *Advanced information technology to counter biological terrorism.* SIGBIO Newsl., **20:2**, 2000, 2–7.

[20] R. Chadha, H. Cheng, Y.-H. Cheng, J. Chiang, A. Ghetie, G. Levin, H. Tanna, *Policy-Based Mobile Ad Hoc Network Management.* POLICY'04: Proceedings of the Fifth IEEE International Workshop on Policies for Distributed Systems and Networks, Yorktown Heights, New York, USA, June 7-9, 2004, 35–44.

[21] M. Hauge, O. Kure, *Multicast Service Availability in a Hybrid 3G-cellular and Ad Hoc Network.* International Workshop on Wireless Ad-Hoc Networks, 2004.

[22] C. Kappler, P. Mendes, C. Prehofer, P. Poyhonen, D. Zhou, *A Framework for Self-organized Network Composition.* Proc. of the 1st IFIP International Workshop on Autonomic Communication, 2004.

[23] C. E. Perkins, P. Bhagwat, *Highly dynamic Destination-Sequenced Distance-Vector routing (DSDV) for mobile computers*. ACM SIGCOMM Computer Communication Review, **24:4**, 1994, 234–244.

[24] S. van der Meer, S. Arbanowski, T. Magedanz, *An Approach for a 4th Generation Messaging System*. Proc. of 4th International Symposium on Autonomous Decentralized Systems (ISADS'99), Tokyo, Japan, March 21-23, 1999, 156–163.

Acknowledgment

The authors acknowledge the EU PalCom:Palpable Computing (IST-002057) FET project, which contributed funding toward development of the RASCAL project. This acknowledgment includes those PalCom consortium members that contributed toward this work including David Svensson from Lund University, Sweden, and Jacob Mahler-Andersen, Jacob Frølund, Henrik Gammelmark and Michael Christensen from Aarhus University, Denmark. We would also like to thanks our colleagues at Whitestein Technologies, Monique Calisti, Giovanni Rimassa, Thomas Lozza and Stefan Thurnherr for their contributions to the work.

Dominic Greenwood
Whitestein Technologies AG
Zürich, Switzerland
e-mail: {dgr@whitestein.com

Roberto Ghizzioli
Whitestein Technologies AG
Zürich, Switzerland
e-mail: rgh@whitestein.com

Whitestein Series in Software Agent Technologies, 81–100
© 2007 Birkhäuser Verlag Basel/Switzerland

Autonomic Resource Regulation in IP Military Networks: A Situatedness Based Knowledge Plane

Gérard Nguengang, Thomas Bullot, Dominique Gaiti, Louis Hugues and Guy Pujolle

Abstract. During the last decade, networks have been growing dramatically on several dimensions: the number of users has been multiplied by more than 15, the amount of traffic by more than 100 and the number of network technologies and usages has also been diversified. One of the main consequences of this growth is the increase in dynamicity of the network behaviour. Therefore, it is more and more difficult for human operators to manage the network in an efficient way. That is why current research in the field of network management tends to automate network control and management. A large initiative has been proposed by IBM to give more autonomy to computing systems: Autonomic Computing Initiative. The paradigm of Autonomic Networking was introduced as an extension of Autonomic Computing, focusing on the autonomy of networks. At the centre of it lies knowledge management. In fact, to be self-managed, a network element needs to know a lot about its environment. But sharing a large amount of knowledge over a large network is very expensive in terms of resources. Therefore, new knowledge management mechanisms are required. A good example of networks that need to be autonomic is military networks. In fact, on one hand, military communication environment is very dynamic, and very uncertain; and on the other hand, quick decisions have to be triggered to guarantee -whenever it is possible- the connectivity for strategical communications. This paper proposes an interpretation of the Autonomic Networking paradigm, a description of a situatedness-based knowledge plane, and an instantiation of these concepts to a concrete application: resource regulation in IP military networks.

Keywords. autonomic, multi-agent, knowledge plane, resource regulation.

1. Introduction

Since several years, communication networks are integrating more and more new services. Development of these new services is made possible by new control mechanisms (traffic control, QoS architectures, differentiation of services, etc.). These new mechanisms induced new layers of complexity. Indeed, every time a control mechanism is implemented, a number of logistic activities are to be performed, including configuration, optimisation, healing and protecting. Moreover, because networks tend to be more and more dynamic, these activities should be performed permanently. This requires intensive care from human operators, which drastically increases the network operation cost. Therefore, a critical perspective is to make the network more self-managing, i.e. adapting itself automatically to the current conditions of traffic and available resources, with less human intervention. To achieve this self-management, several approaches were proposed by different research teams. In 1996, [1] proposed a multi-agent based distributed intelligent ATM network management architecture, which enabled the ATM network to manage itself. In 2001, IBM raised a large research impulsion with their Autonomic Computing Initiative [2]. In each of these cases, the new mechanisms/architectures that are developed to make the network self-managing require a huge amount of symbolic and/or numeric knowledge. Moreover, these pieces of knowledge are very heterogeneous in their aggregation/abstraction level and in their nature, and this heterogeneity may cause many difficulties, in particular coherence, pertinence and redundancy problems in the gathered knowledge. To address these issues, a new plane extending the standard 3-plane architecture was proposed: the knowledge plane. This new plane would assemble a range of mechanisms which gather, compute, exchange and provide to the network elements all of the knowledge they could need (including for example control information and management information). Because of their changing and uncertain environment, and because of their need for fast response, military networks are a good example of networks that need to be autonomic. In fact, since the environment can change very quickly, it is hardly possible to reconfigure the network manually to enable reliability for strategical communications. On the contrary, this reconfiguration has to be done automatically by the network itself.

This paper presents an implementation of the knowledge plane based on a situated view, and uses this plane to control and regulate the network resource in a military context. The first part of this paper draws the borderlines of autonomy in computing and describes the two main approaches to autonomy: what is autonomous and what is autonomic. In the second part, we present the knowledge plane and the different approaches and points of view on this new plane. The third part presents the way we use multi-agent systems to control network environments. The concept of situatedness is introduced here, and we discuss the meaning of neighbourhood in a network control context. The last part presents our implementation of a situatedness based knowledge plane. An instantiation of this knowledge plane in a dynamic military communication environment is then proposed.

2. An Overview of Autonomy in Computing and Networking

The concept of autonomy in computing was introduced two decades ago by Artificial Intelligence (AI) research teams under the topic of Autonomous Agents. The focus was set on the capability for software agents to achieve goals without the guidance of human operators. A common application of such systems was robotics. Robots have to behave in such a way that they can achieve on their own the goals they were designed for. A large research field associated with autonomous agents is action selection (What to do next?). This involves sensing the environment, analysing it, planning and scheduling of the actions to achieve goals considering the given environment and acting on the environment. An agent is told to be autonomous or not regarding its capability to achieve its goals without human intervention.

At the same time, some research teams worked on agent survivability. Agent survivability is the capability for an agent to take care of itself, in order to preserve its own integrity and optimise its performance. Indeed, in many cases, it is much more useful for a system to adapt, protect, optimise itself, than to make plans to achieve a given goal. In 2001, following this latter preoccupation, IBM released a large research and business initiative, called Autonomic Computing. Though the concepts and methods behind are very close to those which were developed under the Autonomous Agents topic, there is a notable difference: Autonomic Computing is entirely focused on systems self-management. The autonomic word refers to the human autonomic nervous system, whose role is to manage unconscious human body activities, like heart beating and inner temperature adjustment. One can remark that in the human body, it is not enough to keep the temperature stable and the heartbeats regular. Depending on the context, these parameters have to be adapted, in order to support the human conscious activities, safely and in an optimal way. This is exactly what autonomic systems have to achieve. More precisely, IBM proposed 4 activities for autonomic systems. To be autonomic, a system will have to:

- Configure itself, which requires to have a read and write access on its own parameters.
- Optimise itself. Using the self-configuration capability, the system will be able to optimise itself.
- Protect itself. This is the ability of detecting attacks as early as possible, and performing adequate defence activities in order to avoid being assaulted.
- Heal itself, which is the first thing to do whenever the system has encountered a failure. This failure can be due to an accident as well as an attack.

2.1. Autonomic computing architecture

To achieve autonomicity in computing systems, the Autonomic Computing Initiative proposes an architecture [3] with several components including mainly a standardised resource management interface and an autonomic manager. Zero, one or more autonomic managers are responsible for one or more resources. An autonomic manager manages a resource using its standardised management interface. The autonomic manager is a software entity which can either be embedded in a physical resource, semi-centralised or centralised in dedicated management systems. A resource can either be a physical system (Database server system for example), a software entity (a Database server, a web server, an applicative server), or a set of physical or software elements. The structure of an autonomic manager is based on a usual AI control loop (Figure 1). The control loop is composed of four tasks: perceiving the state of the managed resource environment thanks to dedicated sensors (through the standard management interface), analysing the situation of the resource in the environment, planning tasks to stay or return to a stable state of equilibrium and executing these tasks thanks to dedicated actuators (through the standard management interface). In the centre of this loop lies knowledge. It is assumed that to achieve self-management in computing systems, provisioning and using knowledge will be a key issue.

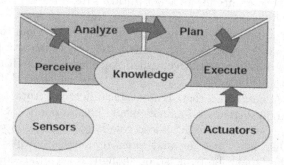

FIGURE 1. Autonomic control loop.

2.2. Autonomic networking

The Autonomic Computing Initiative does not focus on large networks of homogeneous routers, but rather on dedicated information systems, made out of heterogeneous components, like Database servers, web servers, application servers, workstations, PABX, and so on. Thus, the complexity of global knowledge management is reduced. Indeed, each piece of equipment has its own specific knowledge, and knowledge sharing between elements is limited. Also, the range of available parameters and possible actions is very large. In the world of homogeneous networks (backbones, ad hoc networks, mesh networks, sensor networks) issues are very specific, and require specific responses. For example, the analysis and plan tasks of the control loop have to stay very simple, because the amount of available resources is rather small. Moreover, the number of available parameters and possible actions on each network element is very limited, which limits the possible complexity of the analysis and plan tasks. Another layer of complexity is involved by the fact that the actions must be coherent over several network elements (if not all of them), which increases the importance of collaboration, and therefore of knowledge sharing, over the network. The key of autonomy in homogeneous networks, thus, lies in knowledge sharing and coordination of network elements, rather than on the ability of taking complex decisions based on complex analyses.

3. The Knowledge Plane

3.1. Different visions on the knowledge plane

Often in the past years, the knowledge plane has been presented as a meta-control plane (i.e., algorithms controlling usual control algorithms). This idea could sound very seducing, because in one single shot, it would address the whole autonomic networking field: managing high level knowledge and using it to enable network self-management. This approach was pushed by Clark *et al.* [4], who propose the knowledge plane to gather information from the network and reconfigure the network in an autonomic way, considering high level goals set by administrators. In a rather close approach, reference [5] proposes an Information Plane, which is designed as a symbolic algorithmic plane. The goal of this plane is to use abstract information about network control and let declarative programs reconfigure network elements according to high level operator policies. We argue that knowledge management has become such a critical field in network management, and particularly for autonomic networking, that it deserves a field by itself. Indeed, separating knowledge management and autonomic control makes it possible to develop new algorithms and architectures for both. Therefore, our approach is strictly different: we think that knowledge plane should only be a protocols/algorithms layer to manage knowledge within the network, in order to design (later) new autonomic control/management algorithms. Thus, our knowledge plane has 3 main purposes:

- Gathering useful information from usual control algorithms (network interface connectivity, load, security, routing information). As often as possible, the knowledge plane should be embedded within network elements themselves, so that this information gathering activity can be performed permanently, with little impact on network performance. When this is not possible, information can be gathered in an external box, using SNMP protocol.
- Giving a meaning to the data gathered, correlating them with each other, formatting them and maintaining a generic instance of them. This step converts the whole gathered information into knowledge. The main difference between information and knowledge is that knowledge is in context. While information is composed of raw data, knowledge is a set of interrelated data, in context, compared to reference values taken from different places at different times.

- Providing knowledge where it shall be useful, in a format that matches the needs and understanding capabilities of the recipients. Numerous mechanisms and protocols have been proposed to match the heterogeneity of knowledge and the heterogeneity of its use. Three kinds of models were proposed then: global-broadcasting models, situated-diffusion models and peer-to-peer negotiation models. An optimal solution stands in a hybrid generic model which allows one to use the benefits of each approach.

3.2. Some words on information models / knowledge models

Reference [6] introduces different styles of information models to find upper and lower bounds to routing algorithms efficiency. They consider separately No-Information models, Complete-Information models and Partial-Information models. Partiality of their information models concerns the precision of this information, rather than its localisation. What they call a partial information model is an information model where decision points have got imprecise information about the whole network. They discuss the need for a large amount of information and propose to use a partial information model.

Reference [7] proposes a classification of knowledge models to address the load sharing issue in networks of workstations. Authors experiment a range of load sharing algorithms which use different kinds of knowledge models. They split knowledge models into two main categories: partial knowledge on one hand and global knowledge on the other hand, which in turn is also split into distributed information global knowledge models and centralised information global knowledge models. As we will discuss later, in such taxonomy, our approach would definitely lie on the distributed information partial knowledge category.

3.3. Integration of a knowledge plane

Figure 2 shows the standard communication model between network elements without any knowledge plane. Each control mechanism has to get information for its own use from the equipment (by executing specific code or polling the nodes neighbours if necessary) and each control mechanism does exchange control information with equivalent mechanisms in other network elements. Thus, it is not possible to avoid redundancy between different control mechanisms. Moreover, it is difficult to design global information management policies. For example, each control algorithm surely needs to know if direct topologic neighbours are alive or not. This kind of information cannot be shared by different mechanisms in such architecture. In the same way, many advanced control algorithms need to know (or would benefit knowing) about the load of each interface. Until now, each control mechanism has to query its own information from the equipment, represent it and use it, all by itself. Though it used to be good enough in usual network control planes, this is definitely not a good knowledge management architecture. Since knowledge management is going to be a key issue for the future autonomic network, a novel global knowledge management design seems unavoidable.

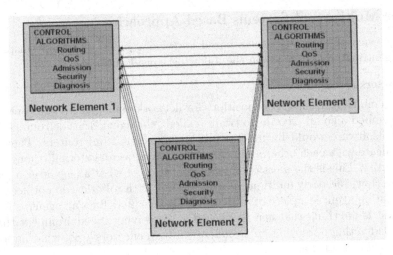

FIGURE 2. Communication between network elements without a Knowledge Plane.

In our vision (Figure 3), the knowledge plane is the only one responsible for information gathering and sharing. Each control mechanism then gets its useful information from it. To provide useful, rich and pertinent information to the different mechanisms, the knowledge plane has to pre-compute data, correlate them and maintain them within a specific rich format. This format may include a confidence level, a creation date, a last modification date, a life time, an info source, etc. Such enriched information can be called knowledge. Knowledge sharing is ruled by knowledge sharing policies which could be smarter than either a point to point negotiation or a global broadcast.

The knowledge plane helps in the mutualisation of the knowledge. It allows different mechanisms in a single network element to share common knowledge. This mutualisation dimension is called "vertical mutualisation". The knowledge plane also allows different network elements to share common knowledge. It is the "horizontal mutualisation". Until now, "vertical mutualisation" does not exist at all, while "horizontal mutualisation" is achieved by each control mechanism on its own through dedicated communication protocols. Paper [8] describes three examples of knowledge plane architectures: a set of agents solving problems by cooperating with each other; a Policy-Based architecture with local PDPs; a unique supervisor that knows all about the network. The authors then choose to base their knowledge plane on a multi-agent system. As we explain in the next section, though we chose to base our knowledge plane on a multi-agent system, our approach is slightly different from the approach proposed in [8].

4. Our Multi-Agent Systems Based Approach

Our approach of the situatedness based knowledge plane is strongly inspired by multi-agent systems. It relies on the following considerations:

4.1. Routers are agents

Common multi-agent based propositions for network management and control propose to embed a small piece of software (called "the agent") inside routers. These pieces of software would be functionally independent from routers. They could communicate with each other and control the router as an external (though collocated) entity. This is the most common acceptation of what an agent is: a piece of software. However, early multi-agent works, along with robotics, do not necessarily assert that an agent is a piece of software. Reference [9] defines an agent as follows: "An agent is anything that can be viewed as perceiving its environment through sensors and acting upon that environment through effectors". So, what happens if we try to see a router as an agent?

4.1.1. A router perceives its environment. A lot of network control mechanisms are based on the information a router can gather from its interfaces. It can perceive the data throughput on each interface; it can perceive events, like link-related alarms or reservation requests; it can access to a number of statistical data concerning the traffic; etc.

4.1.2. A router acts on its environment. In the same way, it is a fact that routers can act upon their environment. Every control algorithm involves acting upon the routers environment. A router can route packets; a router can slow down

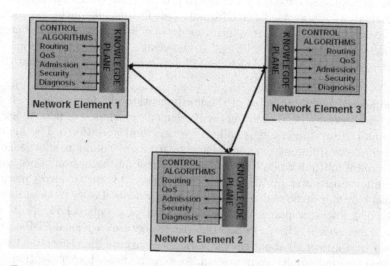

FIGURE 3. Communication between network elements with Knowledge Plane.

flows; a router can drop packets; etc. So yes, definitely, a router can act upon its environment. If a router can perceive its environment, and act upon it, then, according to [9], we can assert that a router is an agent.

4.2. The network is a multi-agent system

Reference [10] defines a multi-agent system as a system "composed of multiple interacting computing elements, known as agents". The keyword of this definition is "interacting". Are routers interacting within a network? Every network control protocol involves communication and coordinated action between routers. Routers are interacting whenever they exchange control messages, information messages and requests. Thus, a network of routers can easily be assimilated to a multi-agent system. Therefore, we can look at each proposition, each new idea in the field of multi-agent systems and search for compatibility with the network routers multi-agent system. Better, we can check if the propositions that were made (and are in use) in the field of multi-agent systems cannot help in some of our network issues. And this is done without adding a software agent within routers. The main characteristic we are interested in is situatedness, as a mean to address the critical issue of scalability of network control mechanisms.

5. Situatedness Paradigm

Situatedness is a knowledge/communication paradigm widely used in the multi-agent systems area. Reference [11] defines situatedness as follows: "Situatedness [...] means that the agent receives sensory input from its environment and can perform actions which change the environment in some way. Examples of environments in which agents may be situated include the physical world or the Internet. Such situatedness may be contrasted with the notion of disembodied intelligence that is often found in expert systems". By extension, this means that each agent possesses its own "personal" vision of the environment, not necessarily the same as its neighbour's one. Agents then communicate to share a part of their knowledge. Each agent knows about its own state and about its neighbour's.

The situatedness paradigm has been largely used for robotic researches. A robot can either base its actions on a global map of its environment, which is a global piece of information or on sensors (light sensors, noise sensors, odor sensors) providing situated information. Robots based on global information behave better than situated ones in a static and well-described environment. On the contrary, situated robots behave much better in a dynamic and/or complex environment. For example, let's consider a robot whose purpose is to search for cheese in a flat. If the flat is perfectly described, the cheese is static and the map up to date, a robot based on global information is much more adapted to this problem. But if the cheese moves, it has no chance to find it. On the contrary, a robot based on an odor sensor has a very high probability to find it. We perceive the interest of such a paradigm, especially in very dynamic and not well described environments. The key word in this paradigm is "neighbourhood". If we think an agent can base its decisions on the perception of its environment, the important point is to know what "its environment" means. Since (as it has been explained in section 4) a router is an agent, one can think that its neighbours are the routers which are directly connected to it. But it is also possible to consider the routers that are 2 or 3 "hops" away or even those that are functionally connected to it, i.e., routers which are interested about its state and environment. Moreover, the idea of a node neighbourhood can be thought as a variable notion. For instance, a node which is a router neighbourhood at a moment can exit from it few seconds after. More generally, the situatedness of an agent can be defined with two main characteristics:

- Type of the situatedness: this is the kind of neighbourhood we consider. A router can either be situated within a geographical area close to the node under study, or within the functional environment of this node.
- Shape of the situatedness: is the neighbourhood static or not, does it include all neighbours in an area, or are neighbours integrated in the situated view depending on other parameters (ease of access, cost, etc.)?

This is in our opinion a key challenge for tomorrow's autonomic networks: everywhere "neighbourhood" makes sense, it has to be used. This "situatedness" is a key to scalability, robustness, and adaptability in highly dynamic large networks.

6. Benefits of a Situatedness Based Approach

Each piece of knowledge shared on the network has a cost. This cost can be seen as the sum of 3 elementary costs: computational load overhead, network load overhead and storage overhead. Computational load overhead represents for each router the cost of processing knowledge pieces that have been received per time unit. Network load overhead represents for each network interface in the network the cost of sending packets containing the knowledge to be shared. Storage overhead represents for each router the cost of storing the knowledge received in its limited storage resources. By doing a reasonable approximation, we can assume that each of these 3 costs (and thus the sum of these costs) is proportional to the number of knowledge pieces received by each router per time unit. Therefore, we propose a simple calculus to appraise the evolution of this value depending on the knowledge diffusion radius.

6.1. Number of Knowledge Pieces (NKP) Per Time Unit

Our purpose here is to propose a simple calculus to appraise the evolution of the NKP depending on the knowledge diffusion radius. In order to find a satisfying approximation for this theoretical value, we assume that the network topology is acyclic and infinite, with each router having the same number of interfaces. We evaluate the number of knowledge pieces a node has to take into account at any time. We call a piece of knowledge an instant data representing one control data of a given network interface. The NKP value depends on two factors: the max distance where information is shared (N_HOPS) and the number of interfaces per node within the topology (N_IF).

- If the node does not use control information at all, even not on its own interfaces.

$$NKP(N\_IF, null) = 0. \tag{6.1}$$

- If the node uses its own interfaces control information: the NKP is 1 for each of its interfaces.

$$NKP(N\_IF, 0) = N\_IF. \tag{6.2}$$

- If the node uses its own interfaces control information, and its direct neighbours interface, NKP equals 1 for each of its own interfaces plus 1 for each of its neighbours ones.

$$NKP(N\_IF, 1) = N\_IF + N\_IF^2. \tag{6.3}$$

A generalisation of this calculus is given by the following relation (Figure 4 gives a representation of this system with N_IF=3):

$$NKP(N\_IF, n) = NKP(N\_IF, n-1) + N\_IF^2 * (N\_IF - 1)^{(n-1)} \tag{6.4}$$

with n=N_HOPS.

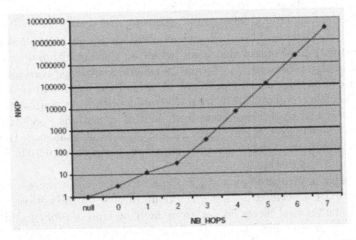

FIGURE 4. A representation of an upper bound of the NKP growth depending on N_HOPS with N_IF=3.

6.2. Discussion on these data

What clearly appears here is not surprising: the cost for sharing knowledge into a network grows exponentially with the distance where each data are broadcasted. However, this leads us to an interesting observation: if we assume that the cost of knowledge sharing is proportional to the NKP, the cost for sharing knowledge in a reduced scope is very small. As an example, in a network where the most distant nodes are separated by 7 hops, the cost for broadcasting information over the whole network would be about 1000 times the cost for broadcasting information to a 4 hops radius area. Therefore, to develop more autonomic control algorithms working with a large quantity of knowledge, and if local knowledge brings enough information, it would be much more efficient to use "wherever it is possible" situatedness based knowledge sharing schemes.

7. The Situatedness Based Knowledge Plane

We propose the knowledge plane to be based on a multi-agent system, assuming that each router can be assimilated to an agent. Each agent builds a primitive situated view of its environment by gathering control data from its hardware layer. In a "multi-agent" vocabulary, the agent sets sensors on each interface in order to sense variations of different parameters. Then, by exchanging sporadic knowledge messages with its nearest neighbours, the agent begins to extend this view. For a range of applications, this situated view can be coupled with global information such as the static topology with routing metrics. In future works, we will try to classify and develop a range of knowledge pieces our knowledge plane could manage. This classification would surely include knowledge pieces related to QoS (QoS parameters, including delay, jitter, and loss) and security.

Until now, we have developed a generic knowledge gathering mechanism to be combined with any control algorithm (whatever this algorithm does control): a generic knowledge representation format, a generic knowledge correlation toolkit and a generic knowledge sharing protocol. First of all, an event management mechanism enables the communication with control algorithms: a simple loop, SNMP or syslog server can read information from control algorithms (load of interfaces, attacks, etc.) and trigger events every time a significant change in the network element state happens. Events are then parsed, in order to enrich the knowledge plane whenever it is justified.

Knowledge is presented as a stack of facets. A facet is a representation of a part of knowledge associated with a given point of view on the network. It is very important to notice that facets do not group facts by kind of problem, but by kind of knowledge. For example, "the state of the links" can be gathered in one facet. This facet can be used for routing, security or healing purposes. On the contrary, routing information cannot be a facet. Facets are presented as lists of facts. A fact is a list itself, containing whatever is useful regarding this fact, plus a time stamp. A list can contain any type of data, including strings, numeric values or even lists. Knowledge correlation toolkit is a set of tools to link, filter, match, add, remove, modify or replace facts within the list. They enable network programmers to very easily develop knowledge management algorithms, using for example rule-based programming (IF (eventA ON nodeB AT timeT1) AND (eventA ON nodeC AT timeT2) AND (timeT1 NEAR timeT2) THEN add fact (superEventA + ON nodes A and C AT timeT1)). Knowledge sharing is done as follows: when an event is detected, the event parsing loop can in the same time add a fact to any local knowledge facet, but it can also decide to diffuse the fact to neighbours. It is then possible to give a diffusion mode, a diffusion radius. The message is sent to corresponding neighbours, depending on the actual topology.

8. Autonomic Resource Regulation in Military IP Networks

Military networks consist of heterogeneous data transmission technologies associating fixed, wireless and mobile infrastructures. The overall network connectivity is volatile. The wireless communication links often suffer from signal attenuation due to Doppler effects and spatial fading, as well as vulnerability to weather, electromagnetic interference, interception and jamming [12]. This results in a high fluctuation of the available bandwidth, high latency (delay) and high error rate. These extremely unfavourable conditions degrade the capability to run network applications. Therefore, the notion of end-to-end quality of service is crucial since these networks are often used for the transmission of mission critical information with specific performance requirements between strategic and tactical domains. However, the conventional Quality of Service approaches mostly based on resource allocation are unable to deal with the high dynamicity of military networks. Especially, the scalable and suitable Differentiated Services (DiffServ) framework [13] which introduces multi-level network services is by essence unable to cope with the unpredictable and unstable nature of these networks. Although, several research were undertaken during the last decade to integrate adaptiveness over the DiffServ architecture [14, 15], the focus was mostly on the automation of the network configuration process and the dynamic allocation of network resource. Indeed, even if sufficient network resource is allocated to each ingress flow, an unforeseeable external event can drastically reduce the available bandwidth and compromise the transmission of both essential and non-essential military traffic.

To overcome this problem, the network should be able to efficiently perform on its own and attempt to sustain as much as possible an acceptable level of service for at least critical data when only a part of all resources may be available. Since the network performance along a path can diminish at a moment and regain its former level a few seconds later, the solution resides in the construction of an environment-aware system that can autonomously pilot the network. This section instantiates the situatedness agents based knowledge plane approach for self resource regulation in a DiffServ based military IP network. Software agents are embedded in edge nodes of the network. These agents continuously update their local perception of the network state and cooperate efficiently in order to build a partial but consistent knowledge on what is happening on the network. Based on this knowledge, the agents react on their networking features in a way that the performance anomaly will be annihilated.

8.1. The proposal overview

As underlined previously, commonly used techniques for dynamic resource management in DiffServ networks are unable to fulfil the military QoS requirements.

G. Nguengang, T. Bullot, D. Gaiti et al.

Since the bandwidth often fluctuates, even higher priority flows can encounter congestion. Such a situation can easily compromise the achievement of a military mission. Therefore, continuous performance monitoring is crucial to tract the on-going QoS, compare the monitored QoS with the expected performance, detect possible QoS degradation and then take on the fly corrective actions accordingly to sustain as much as possible the required QoS for at least some flows and avoid the collapse of the service.

Our approach to solve the aforementioned problem is to build an environment-aware self-regulating distributed system that pilots the network according to the data in the knowledge plane. Since the network core is in some way incontrollable, our proposal consists in the implementation of the distributed knowledge plane agents in the edge routers of a military domain. The network core is then considered as the environment the so formed multi-agents system has to deal with. (Figure 5). At a bootstrap stage, each agent is configured by the Network Opera-

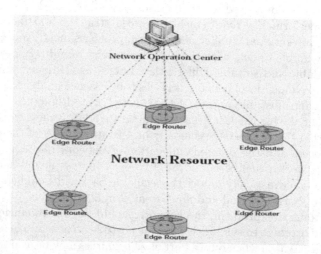

FIGURE 5. Overview of the proposal.

tor Centre (NOC) and receives a file made of incoming network flow descriptions with the associated network services, the description of the QoS requirements for each service and the probing traffic characteristics. This information constitutes the Agent goal Specification (AgS). The Agent goal Specification drives the agent perception of the network state and provides the context for the knowledge construction. Once the system is settled, each agent probes periodically its predefined network services searching for any performance anomaly, builds and consolidates its knowledge from peers information and acts when necessary (based on its knowledge) on the network by suspending or upgrading incoming flows. In short, each agent carries out two behaviours: the resource diagnostic behaviour and the resource management behaviour.

8.1.1. The agent Resource Diagnostic Behaviour (RDB). The RDB allows the active monitoring of all the defined network services on an edge to edge basis. The agent situated view (Figure 6) is therefore made of all the paths connecting it with the edge routers defined by the network operator in the AgS. For each service, the probe generates micro test traffic in direction of all the nodes of its neighbourhood (see the section on Situatedness paradigm) and computes the QoS metrics [16, 17].

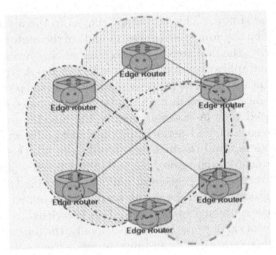

FIGURE 6. Situated view representation.

In this work, with an aim of simplification, we decide to restrict the QoS metrics to the end-to-end delay between edge routers. Once that the edge-to-edges delays are measured for each service, a local indicator per service is then computed. The indicator varies between 0 and 1. The computing process of the local indicator iS for service S is as follows:

- *Step one: Selection of the highest delay value for the service S* e.g., measured delay D, from Agent A to Agent B = (D,A,B) where D is the highest delay measured for the service S.
- *Step two: Indicator computation*

$$iS = \begin{cases} \frac{Measured\_Delay}{Max\_Authorised\_Delay} & \text{if} \quad Measured\_Delay \leq Max\_Authorised\_Delay \\ iS = 1 & \text{if not} \end{cases}$$

When the indicator iS is less than 1, nothing is wrong on service S. Oppositely, when iS = 1, it implies that the flows using the service are in trouble. Something must be done to rectify the situation. The computed indicators are stored in the knowledge plane and the (iS,A,B) style events are broadcasted by the agent to its neighbours.

8.1.2. The agent Resource Management Behaviour (RMB). The RMB consists of the decision-making process and the execution of the adequate actions in order to adapt the network traffic according to the available network resources and avoid the complete collapse of network services. The RMB takes as input the set of indicators available in the router knowledge plane and checks for each service if it is necessary to initiate a coordinated action. The possible actions are either degrading flows to release the congested resource or restoring the initial PHB of the previous degraded flows. This is realised thanks to the agent control module which acts directly on the admission control module of the router operating system.

For each service, the first step of the decision making process is to determine the edge router that appears more than the others in the destination field of the indicators. This indicates that the concerned destination is in serious trouble. If the greatest of these indicators exceeds a certain threshold, a message is sent to all the agents of the situated view. A decision must be taken to release the resource and avoid persistent congestion and performance degradation. It will be taken by one of the agents having received the message according to whether it is the smallest in the collating sequence, each agent being identified by an alphabetic letter.

Figure 7 summarises the process of building and using the knowledge. The measurements realised by the probe are associated to a context to become knowledge thanks to data contained in the AgS. This knowledge is completed by the environmental perception of the other agents. Finally, the decision making process of the agent uses this knowledge as an input to guide the undertaken actions.

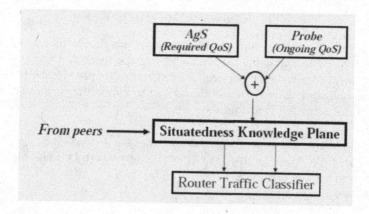

FIGURE 7. The process of bulding and using the knowledge.

9. Experimental Setup: Video streaming in dynamic environment

For validation purpose, we have chosen to consider a video streaming application in a context where network resource is variable as in a real military context. Our choice was guided by the fact that it is very easy, just by watching a screen, to evaluate the relevance of agents' actions on the system. This test can be replayed with others delay sensitive streams such as voice. A test-bed has been carried out to test our concepts and evaluate the relevance of our approach. Figure 8 shows the detailed setup. It consists of: 4 PC-based Linux routers, 2 video servers and 2 video renderers. The VideoLAN software is used for the video streaming and viewing.

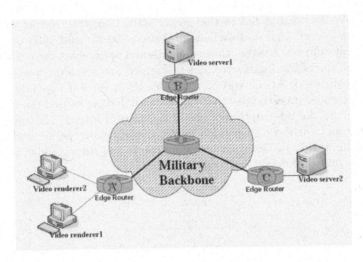

FIGURE 8. The test bed.

To simplify the tests, we considered that the network provides only one service, the video service. This represents the situation where all traffic classes are overloaded and it is impossible to borrow bandwidth. The authorised maximum delay for the video service is fixed at 500 ms. The agents are implemented in Java and the probe module in C. The test scenario is as follows: 2 video flows are streamed in loop from the servers to the renderers over the UDP protocol (Figure 8). Each agent is configured with the corresponding AgS. Thus on each screen, it is possible to see two different videos. In a first stage, sufficient network resource is provided. All the links are point-to-point 100Mbit Ethernet. The videos have a perfect viewing quality. No degradation is observed on the images. Then, to emulate a performance problem in the backbone (perturbation on satellite communication for instance), the bandwidth of the link which connects the video renderers' edge router to the backbone is reduced to 7.6Mbit. This is done thanks to the token bucket filter queue discipline of the Linux traffic controller module. With this bandwidth, the streams start to experience degradations. The video service is impacted by the lack of bandwidth. The images become fuzzy on the screens. The multi agent system is not yet activated. So, if nothing is done to reestablish the initial status, all the end-users video applications will suffer of this performance degradation. Under these conditions, the agents are activated within the edge routers. The lack of sufficient network resources is detected by the system and an agent is elected to prohibit randomly the access to the network of one of its incoming flows. This results in the improvement of video service delivery and a good visual quality for the remaining streams. In a multiservice context, the flow prohibition can be replaced by the attribution of a low priority to the flow. Figure 9 shows the delay variations of the video service between the video servers edge routers and the customers edge router few seconds before the agents activation and after. We notice that before the agents activation, the transfer delay of the video service is very high and oscillates between 700 ms and 1000 ms. Once the multi-agent system is activated, this delay falls and approaches zero. This is due to the fact that one of the ingress streams is stopped and the shared resource is not any longer congested. At the end of a variable time, since it depends on the network state, the system estimates that the anomaly having caused the performance degradation and the interruption of one of the flows is solved. The agents cooperate together and authorise suspended flows to reach again the network. Since the bandwidth limitation still exists, the system reacts and stops a video stream. This explains the jumps of delay observed thereafter.

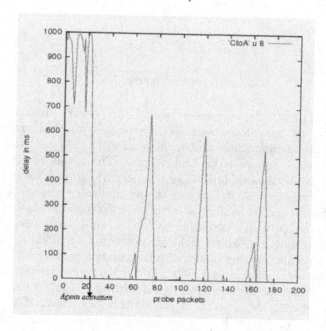

FIGURE 9. Network delay variation.

10. Conclusions

This paper presents our vision on autonomic networking. We explain that achieving this goal will require new control algorithms which will take into account a larger range of situations, to address more and more network issues. These new advanced algorithms will definitely need different and unified knowledge management policies. The Knowledge Plane, thus, would be an independent software area where new knowledge management mechanisms would gather, compute, exchange and provide formalised pieces of knowledge to advanced control algorithms. This activity can be split into two main roles: vertical mutualisation and horizontal mutualisation.

As an implementation of this Knowledge Plane, we adapted the situatedness paradigm (which is used by Artificial Intelligence / Multi-Agent research labs) to a network context. Thus, we developed a situated view of control information. This situated view can gather for example load information, connectivity information and most used destination addresses of near neighbours. We proposed a definition of what neighbours should mean in our situated view and gave examples of how it could be implemented.

We show in an example that this situatedness based Knowledge Plane will increase feasibility of autonomic control and self-management mechanisms. This example was taken out of a large range of possible applications and is representative of different aspects of the use of a Knowledge Plane.

References

[1] D. Gaiti, G. Pujolle, *Performance management issues in ATM networks: traffic and congestion control.* IEEE/ACM Trans. Netw. 4, 2 (Apr. 1996), 249–257.

[2] IBM Corp., *An architectural blueprint for autonomic computing.* USA, Oct 2004.

[3] J. Koehler, D. Gantenbein, C. Giblin, R. Hauser, *On Autonomic Computing Architectures.* Technical Report, IBM Zurich Research Laboratory 2003.

[4] D. Clark, C. Partridge, J. C. Ramming, J.T. Wroclawski, *A knowledge plane for the internet.* Proceedings of the 2003 Conference on Applications, Technologies, Architectures and Protocols for Computer Communications, August 25-29, 2003, Karlsruhe, Germany.

[5] M. Wawrzoniak, L. Peterson, T. Roscoe, *Sophia: An information plane for Networked systems.* In Proceedings of ACM HotNets-II, November 2003.

[6] M. Kodialam, T. V. Lakshman, *Dynamic Routing of Bandwidth Guaranteed Paths with Restoration.* Proceedings of Infocom, March 2000.

[7] G. Bernard, D. Steve, M. Simatic, *A survey of load sharing in networks of workstations.* Distributed Systems Engineering, 1, 1993.

[8] D. Gaiti, G. Pujolle, M. Salaun, H. Zimmermann, *Autonomous Network equipments.* WAC 2005, Athens, Springer LNCS, 2005.

[9] S. Russell, P. Norvig, *Artificial Intelligence: A Modern Approach 2nd edition.* Prentice Hall, 2002.

[10] M. Wooldridge, *An introduction to multi-agent systems.* John Wiley and Sons, Ltd., Chichester, England, 2002.

[11] R.N. Jennings, K. Sycara, M. Wooldridge, *A Roadmap of Agent Research and Development, Autonomous Agents and Multi-Agent Systems.* Journal on Autonomous Agents and Multi-Agent Systems, **1:1**, pp7–38, 1998.

[12] T. Andrew Au, C. Tran, *Enabling Headquarters Reachback: Adaptation of Collaborative Applications for Tenuous Communications.* Defense Science and Technology Organisation -TR-1588 Report, Autralia, June 2004.

[13] S. Blake, D. Black, M. Carlson, E. Davies, Z. Wang, W. Weiss, *An Architecture for Differentiated Services.* RFC 2475, 1998.

[14] W. Changkun, *Policy-based network management.* In Proc. of the International Conference Communication Technologies, **1**, 2000.

[15] L. Breslau, S. Jamin, S. Shenker, *Comments on the performance of measurement-based admission control algorithms.* In IEEE INFOCOM, 2000.

[16] G. Almes, S. Kalindindi, M. Zekauskas, *A One Way Delay Metric for IPPM.* IETF RFC2679, 1999.

[17] *IPPM-WG, IP Performance Measurements Working Group.* http://www.ietf.org/html.charters/ippm-charter.html.

Gérard Nguengang
Ginkgo-Networks SA / LIP6
104 Avenue du Président Kennedy, 75016 Paris, France
e-mail: **gnguengang@ginkgo-networks.com**

Thomas Bullot
Institut Charles Delaunay/LM2S, CNRS FRE 2848, UTT
12 rue Marie Curie, 10010 Troyes, France
e-mail: **thomas.bullot@utt.fr**

Dominique Gaiti
Institut Charles Delaunay/LM2S, CNRS FRE 2848, UTT
12 rue Marie Curie, 10010 Troyes, France
e-mail: **gaiti@utt.fr**

Louis Hugues
Ginkgo-Networks SA
104 Avenue du Président Kennedy, 75016 Paris, France

Guy Pujolle
Laboratoire Informatique de Paris 6
104 Avenue du Président Kennedy, 75016 Paris, France
e-mail: **guy.pujolle@lip6.fr**

Whitestein Series in Software Agent Technologies, 101–126
© 2007 Birkhäuser Verlag Basel/Switzerland

Autonomic Service Access Management for Next Generation Converged Networks

Monique Calisti, Roberto Ghizzioli and Dominic Greenwood

Abstract. This chapter presents the Living Systems Autonomic Service Access Management Suite, LS/ASAM, a comprehensive middleware solution enabling adaptive connectivity management of nomadic end hosts across heterogeneous access networks with autonomic optimisation of network performance and availability.

1. Introduction

Next Generation Networks, NGN, are becoming increasingly open, shared and with infrastructure that is reliant on highly distributed components. This is largely being driven by the vision of ubiquitous broadband access that is continually evolving the way business and consumer customers interact. These networks must thus continue to improve in terms of performance through multiple dimensions, including for example service mobility, personalisation, transparency and immediacy.

This evolution of network infrastructure offers operators the possibility to create many new forms of business. However, it also poses some significant new challenges in many areas of communications and service management, especially in resource-limited access networks. The NGN view is to rely upon an all-IP infrastructure, offering a clean separation between network and service layers and enabling QoS provisioning "out of the box", which should be easier to manage and less expensive to maintain. However there are several factors which complicate the overall NGN picture.

End users are increasingly demanding new services and dynamic, case-specific service aggregations, to support a seamless and consistent experience across multiple access technologies, devices and locations. They expect to be always best-connected, i.e., to have anywhere and anytime access to the best available technology with the maximum capacity on offer, plus easy-to-use and problem-free services, all at ever lower prices.

Indeed, the proliferation of applications, services and heterogeneous technologies, including advanced multi-modal end users devices, enables a variety of ubiquitous deployment scenarios, but also poses significant challenges in terms of service usability and personalisation. This is further complicated by the need to integrate new solutions with legacy systems, while optimising resource-limited consumption (e.g., radio frequency in access networks).

In addition, the widespread expansion in the availability of high-speed broadband access technologies including cable, DSL, powerline, satellite and wireless, is encouraging the entry of new service providers in both the fixed and mobile telecom sectors, thereby stimulating a competetive environment. In response operators need to identify means of lowering operating costs by optimising service provisioning performance and connectivity management.

It is our belief that a fresh approach is required to achieve these objectives. We thus propose a comprehensive policy-driven, autonomic software solution spanning provider infrastructure and end-user devices that positions auto-adaptive control software directly within the devices. The majority of service and connection provisioning approaches in use today tend to operate on the traditional client/server model and are thus rather ineffective due to a common inability to handle the increasing dynamicity and diversity of heterogeneous access technologies. In this perspective, emerging solutions need to be "autonomic" by design; their components should be able to self-regulate and dynamically optimise their own behaviour according to detected changes in their host environment [1].

We call our approach the Living Systems Autonomic Service Access Management suite, LS/ASAM. It is a comprehensive and innovative solution that enables effective delivery of next-generation ubiquitous services by dynamically combining end user requirements and service provisioning policies with network-facing management and control functionality. By automating selected low-level processes on both the user and operator sides and introducing more "personal intelligence" (user context and behaviour awareness) and "network intelligence" (network services, content and resources awareness) throughout the whole service delivery chain, the LS/ASAM solution realises *Autonomic Service Access Management* (ASAM). The guiding ASAM vision is to use autonomic techniques that enable operators to efficiently manage and optimise resource utilisation, performance and end user experience. This is achieved by transparently tuning service parameters while taking into account changes in both the client and network context.

This chapter continues with a discussion of the ASAM core principles before presenting the architecture and features of the LS/ASAM Suite as a means to realise the ASAM vision. We then describe two key deployment scenarios coupled with a discussion of some of the most distinctive characteristics.

Subsequently we provide some data on experimental work conducted in the laboratory on performance analysis of the LS/ASAM suite prototype. The ASAM simulator is described before the presentation of selected results from recent experiments.

We conclude the paper with some discussion remarks, experimental conclusions and targets for ongoing work.

2. Autonomic Service Access Management

Due to the increasing deployment of multiple access technologies at the edges of networks, the management of ubiquitous communications and services is changing rapidly. Intelligence and specific management and control functions need to be migrated toward the edge of the network and even onto the customers' devices. In particular, *service access management*, i.e., the set of functions including the selection and maintenance of one of several available communication channels, is increasingly demanding:

- Fast and appropriate adjustment of the relevant connectivity parameters to a continuously changing network environment.
- The assurance of sufficient service quality and reliability, whose perception can vary from one user to another.
- In coordination with the aforementioned points, the optimisation of resource usage and reduction of operational costs.

Autonomic Service Access Management, addresses these issues by dynamically and automatically adapting the configuration and utilisation of available network access resources in a reliable and cost-efficient way. This is achieved by embedding specialised intelligence into complex multi-technology and multi-service access networks, including end user devices. The chosen approach is to deploy smart techniques allowing operators to efficiently manage and optimise resource utilisation, performance and end user experience. This is achieved by transparently tuning service parameters (e.g., bandwidth, average delay), while taking into account changes in the context, including user preferences, Service Level Agreements (SLAs), user location, devices features and network resources.

ASAM bases its adaptivity on the capability to autonomously observe, extract, understand and use context information to consequently modify its own functionality. Information exchange and correlation between client devices and access nodes, as well as between access nodes even of different technologies, is at the core of this approach. In particular, through dynamic mediation between (often conflicting) requirements on the client and network side, capacity for given connection requests is allocated by taking into account the status of the whole service provisioning chain. This requires accounting for a variety of parameters that characterise the connection to be created, the consequently required network resources and the policies existing both on the user and provider side.

For this to be realised, flexible and distributed monitoring, configuration and maintenance tools need to be smoothly interfaced and integrated within the evolving networking environment and pre-existing management systems. This is not an easy task, especially when considering that many operators must deal with a diverse mix of systems and processes that make it difficult to effectively monitor and tune service performance once already in the delivery phase. In this perspective, a new kind of management solution is needed. A comprehensive policy-driven and autonomic architecture, spanning basic infrastructures and end-user devices, which builds adaptive control functionality directly into the corresponding elements, enabling the shift of focus from technology to value-added services.

LS/ASAM is a comprehensive ASAM solution that addresses these challenges by making use of software agent technology [2]. Autonomous agents that adapt to changes in the environment, minimising human intervention and service interruption, lie at the foundation of LS/ASAM and provide a powerful means to engineer a distributed and autonomic system that includes:

- Customisable and adaptive routines for automating and tuning repetitive information and control tasks.
- Coordination mechanisms enabling the spontaneous collaboration and dynamic aggregation of services.
- Abstraction of communication components to support context changes through adaptation of semantic grounding.

In this way, autonomous software agents acting as autonomic managers, see Figure 1, are enabling LS/ASAM to exhibit self-management capabilities that increase reliability and performance while reducing operational and management costs. This shifts the burden of many support and control tasks from users to the underlying solution, which assists, facilitates and empowers human decision making.

More specifically, LS/ASAM is a middleware solution empowered with autonomic self-management capabilities, including:

- Self-configuration: policy-based self-configuration of the Suite's components according to changes in their usage and working environment.
- Self-optimisation: proactive monitoring and control of resource usage, performance and end user experience to enforce optimal behaviour.
- Self-healing: automatic fault discovery and correction, both on the end user devices and network elements.
- Self-protection: automatic detection of and protection from unauthorised system control changes.

Control over LS/ASAM components is expressed through policies bound to user preferences and business goals. The system senses, analyses, plans and executes changes in the environment to ensure that business goals can be effectively met.

Although other approaches have been proposed in the literature that address part of the ASAM challenges, none, to our knowledge, is able to dynamically mediate between network and client requirements and accommodate resource allocation and consumption accordingly. In particular, the solution presented in [3], which is the closest one to LS/ASAM, supporting vertical handover in radio access networks. In this system, a dedicated decision module, placed within a concrete provider system, can communicate with various network devices, including client

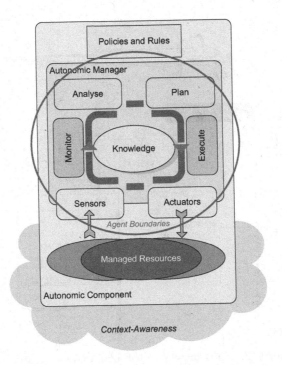

FIGURE 1. An autonomic component architecture.

devices, to determine radio access network selection based on QoS parameters. Some degree of negotiation takes place, but only between entities within the network and excluding the client devices that remain passive.

3. The LS/ASAM Suite Architecture

The LS/ASAM architecture includes two main types of autonomic software components, as depicted in Figure 2, which communicate by relying upon the use of common interaction protocols and a shared semantics-based ontology defining all LS/ASAM concepts. These components are:

- *LS/CA, the Living Systems Connection Agent,* is a client component that can run on a variety of mobile end user devices (e.g., laptops, PDAs, smart phones) and provides mobile users with improved quality and reliability by optimising service access through adaptive connection handover across multiple access technologies and dynamic mediation of service delivery parameters on behalf of the end user.

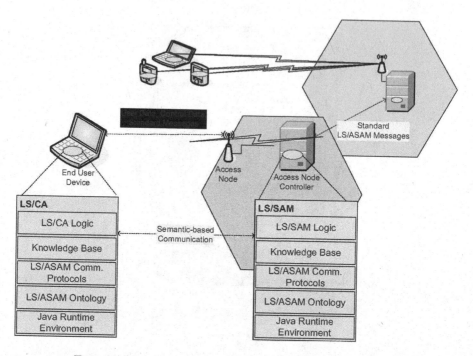

FIGURE 2. An overview of the LS/ASAM architecture.

- *LS/SAM, the Living Systems Service Access Manager*, is a network component that can run on hardware located at the access nodes or at a network management facility. It dynamically optimises resource allocation across heterogeneous network access domains with adaptive problem recovery and load balancing techniques.

  These lightweight software components, i.e., they can live as processes in a Virtual Machine, can flexibly complement and extend many existing service management architectures, and are able to run on resource-limited devices and support asynchronous communication with intermittent network connections. By dynamically coordinating their actions and behaviour, they enable adaptive communication service access by mediating between operator policies and end-users requirements and preferences.

3.1. The Living Systems Connection Agent

The LS/CA component provides adaptive service access by setting connectivity parameters according to the outcome of a mediation process to establish a service access agreement based on the end user's requirements and the network provider's offering. This is determined by a set of factors including:

- Quality requirements of the applications and services running on the device the LS/CA is embedded in.

- Physical end user device status, e.g., battery power level, and properties, e.g., available network interfaces.
- Existing service provisioning conditions according to pre-defined subscription contracts/SLAs.

The LS/CA proactively manages and processes this information according to policies which capture end user preferences, e.g., minimising connection costs, maximising battery life when on-the-move, etc., and supports the following main features:

- *Seamless handover* and *session continuity*. This guarantees interruption-free service access across multiple technologies by allowing an LS/CA empowered device to maintain the same IP address for an entire session. This is achieved by making use of Mobile IP technology [4].
- *Secure communication*. Tight integration of the LS/CA with several third party VPN clients allows permanent secure connectivity. Furthermore, by integrating IPSec [5] and Mobile IP, the LS/CA ensures end-to-end encryption of all generated traffic (as an optional feature).
- *Connection adaptation*. This indicates automatic detection of available networks and selection of the preferred network adapter (access technology) based on service requirements and network conditions for improved reliability and QoS. This can trigger dynamic mediation between the LS/CA and the LS/SAM components.
- *Context-aware user support*. Through semantic service specifications, policy-driven decision making and dynamic information retrieval, the LS/CA improves end-user experience by directly addressing low-level issues (e.g., failure recovery, connection adaptation), while taking into account user policies and boundary constraints, i.e., context-based information and coordination with LS/SAM components as needed.

From the LS/CA perspective the mediation process is initiated by sending a Call For Proposal, CFP, to one or several LS/SAMs. Naturally any LS/SAM with an open connection established with the LS/CA may also receive the CFP so that it can also participate in the connectivity mediation process.

3.2. The Living Systems Service Access Manager

The LS/SAM component proactively monitors traffic and resources in the access node it controls, triggers appropriate actions (e.g., vertical handover, load balancing) according to the network status and current traffic conditions, processes incoming LS/CA calls for proposal and elaborate offers as appropriate - see Section 3.3. In particular, the two main distinctive features enabling LS/SAMs to optimise resource consumption at the access network level are:

- *Load-balancing*. Balancing traffic load across WLAN and cellular networks while considering the QoS needs of running services renders the network more resilient to traffic peaks. This is achieved by dynamic coordination between LS/SAMs that can hand over a certain number of connections to neighbouring access nodes according to possibly several operator policies. The use of distributed constraint satisfaction algorithms [6] for LS/SAMs peer-to-peer orchestration enables effective load balancing by taking into account all existing constraints.

- *Congestion recovery.* Real-time and proactive detection, analysis and relief of congestion, reduces call dropping and increases service resilience and availability. Within an access node, once no new network connection can be accepted or the total requested bandwidth exceeds the total available one, i.e., packets are dropped, an LS/SAM can decide upon specific policies and existing SLAs (if any) whether and how to drop or hand over part of the traffic to neighbouring access nodes.

LS/SAMs decisions and behaviour are guided by the operator's policies that express service provisioning preferences with respect to a variety of aspects including, e.g., how to allocate traffic to balance out network utilisation, how to treat specific users (i.e., connections) in case of congestion, how to adapt pricing schemes according to the user's subscription type. This requires dynamic management of information including:

- Traffic conditions and resources available within the access node the LS/SAM is controlling.
- Traffic conditions and resources available in other access nodes that a given portion of traffic can be handed over to, via dynamic LS/SAM-to-LS/SAM coordination.
- Existing service provisioning conditions according to pre-defined subscription contracts/SLAs.

3.3. Adaptive Coordination of the LS/ASAM Components

The mediation process conducted between the LS/CA and LS/SAM components consists of a sequential interchange formulated as a contract-net protocol [7] negotiation with the goal of determining the best connection parameters given the requirements of the end user, the offering of the network provider and the conditions of the transmission medium.

The requirements of the end user toward the provider are a combination of (i) the preferences of the end user formulated as user policies (e.g., minimising connection cost), (ii) the quality demands of the applications running on the end user device (e.g., a given application may require low end-to-end delay), (iii) the status of end user device resources (e.g., battery power, which can affect the selection of the transmission technology), (iv) the technologies supported by the end user device (e.g., only WLAN and UMTS network interfaces available), and (v) the conditions stated in the subscription contract (e.g., costs for using certain technologies).

The offering of the provider toward the end user is determined by considering (i) the properties of the provider network (e.g., diversity of network access technologies), (ii) the network status (e.g., distribution of traffic load, delay times),

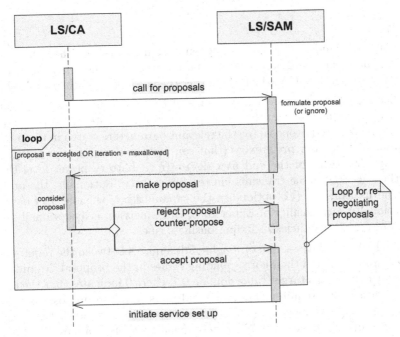

FIGURE 3. Mediation process between the client and network LS/ASAM components.

(iii) the capabilities of the network (e.g., mobility support, QoS control) and (iv) the provider policies, including business rules, that relate to the use of its infrastructure, pricing schemes, traffic prioritisation mechanisms, etc.

Figure 3 illustrates the typical message exchange during a proposal setup sequence. An LS/CA sends a CFP to one or several LS/SAMs requesting offers to set up a connection with specified constraints including quality requirements, or connection characteristics.

An example of a simple CFP is:

```
(set up connection, (min. bandwidth: 100 KBit/s.
max. delay jitter: 50 ms))
```

Once sent to all prospective LS/SAMs, the LS/CA waits until some predefined deadline to receive proposals and/or rejections. Any LS/SAMs that have not sent a proposal or rejection by this deadline are considered to have been unable or unwilling to respond to the CFP. A simple example of a proposal sent by a responding LS/SAM is:

```
(set up connection, (network: UMTS, min. bandwidth: 100 KBit/s, max.
bandwidth: 120 KBit/s, max. delay jitter: 40 ms, max. end-to-end
```

```
delay: 200 ms))
```

This proposal includes some additional connection parameters than those present in the orginal CFP. Although not mandatory to do so, these can be taken into account by the the LS/CA when evaluating the suitability of the proposal.

The proposals are assessed by means of the Proposal Assessment Function (PAF) that takes as input (i) the set of quality requirements stated in the original CFP, (ii) the received proposal (or the relevant parameters stated in the proposal), (iii) optionally, the user preferences (that can be formulated as user policies), (iv) optionally, the status of the end user device (e.g., battery power level that can affect the selection of the transmission technology), (v) optionally, the properties of the end user device, (vi) optionally, the capabilities of the end user device and (vii) optionally, any Quality of Experience, QoE, metrics, (vii) optionally, the set of network operator policies including business rules.

The PAF computes a sum of weighted differences between the required quantitative parameters and their corresponding values in the proposal. Nominally, the PAF is normalised to a target value domain 0,1 where 0 indicates that the proposal does not satisfy any requirements and 1 indicates that the proposal is valid and fully acceptable. Intermediate results between these bounds indicate the degree to which the proposal meets the CFP requirements. Ancillary annotations record if the proposal exceeded the CFP requirements for use with counter-proposal negotiations.

At this point the LS/CA must decide whether to make a counter-proposal to any number of selected LS/SAMs that responded favourably to the original CFP. This decision is made in accordance with how well a received proposal meets or exceeds the original CFP request. If selected, a counter-proposal can be issued to a responding LS/SAM in an attempt to initiate bilateral negotiation to revise the proposed offer. Multiple counter-proposal negotiations can be handled concurrently by an LS/CA with active PAF based comparison of each to determine variances between returned proposal updates thereby assisting with refining individual negotiations by taking into account all ongoing negotiations.

A counter proposal is created by modifying a received proposal in accordance with preferred characteristics. If the original CFP sent was:

```
(set up connection, (min. bandwidth: 100 KBit/s))
```

With a received proposal being:

```
(set up connection, (min. bandwidth: 70 KBit/s))
```

The PAF determines that this received proposal is close to its requirements, as expressed in the original CFP, and thus creates a counter proposal in order to initiate fine-grained bilateral negotiation with the sender of the proposal. The counter proposal in the instance of this example may be that the proposed 70Kbit/s band-width offer is iteratively increased to 80 KBit/s:

Counter proposal: (set up connection, (min. bandwidth: 80 KBit/s))

This counter proposal is a compromise between the original bandwidth spec-ified in the CFP and the bandwidth offered in the returned proposal.

It is important that the decision process exhibits a convergent behaviour to avoid continuous proposal revision. Several suitable algorithms can be found in the literature include that by Hofbauer *et al.* [8] and by Shamma *et al.* [9].

When, or if, a proposal is accepted the client device sends an accept-proposal message to the corresponding network provider. All other proposals that have been received are explicitly rejected by informing their source providers. The reason for rejection may be included in the message.

3.4. Technology Foundation

As networks grow increasingly larger and more complex, they become harder to manage efficiently and reliably. This is even more challenging in resource-limited access networks, which affects the capability to deliver true seamless mobility. Thus, network and service management solutions are required to exhibit autonomic behaviour.

Their components detect, diagnose and repair faults, adapt their configu-ration and optimise their performance, while protecting and healing themselves according to changes in the network and operating environment.

The key idea is to assist, facilitate and empower humans (operators, network administrators, customers) by shifting the burden of many support and control tasks from them to the underlying solution components.

As anticipated in Section 2, the LS/ASAM Suite has been conceived and realised by embedding autonomic self-management capabilities at the core of its functionality. Its components autonomously observe, extract, understand and use context information to consequently modify their functionality, according to poli-cies that are bound to business goals. The autonomic capabilities of the client components, LS/CA, and the network component, LS/SAM, are classified as fol-lows:

Self-configuration. The LS/CA adjusts its own configuration according to changes in the working environment in which the user device is located. Policy-controlled profiles for different locations identify the configuration of features to be used, e.g., connection type, VPN, file shares. The LS/SAM performs self-configuration determining its own behaviour to achieve high-level directives. This enables the network (namely the access resources the LS/SAMs control, e.g., base stations or access points) to respond dynamically to changes in operator policies and/or network state. Different load balancing strategies may be adopted, depending on traffic conditions, resource availability and SLAs.

Self-optimisation. The LS/CA selects a specific connection type according to user policies and in relation to changes in the context. This is particularly beneficial while roaming in partner networks where the nominal connection may not be the preferred, best or indeed cheapest option. The choice of alternative network adapters can also be triggered by the need of optimising specific application performance in relation to device properties and network status. The LS/SAM efficiently manages access node resources to meet specified performance objectives under dynamic operating conditions. By proactively balancing load across distinct access nodes (via interaction with peer LS/SAMs) and triggering vertical handover of selected connections, it is possible to optimise network performance and availability according to existing operators policies.

Self-healing. The LS/CA detects faults in related system components (e.g., network cards, drivers, system interrupts) and transparently takes action to repair and circumvent the anomalous behaviour. The LS/CA also attempts to re-establish lost connections or, if not possible, seamless transitions to a session over an alternative connection type. The LS/SAM is able to detect and repair unpredictable conflicts between service requirements and available network resources. If appropriate, it coordinates its behaviour with other LS/SAMs. In particular, real-time and proactive detection, analysis and relief of congestion allows the LS/SAM to reduce call dropping and thereby increase service resilience and availability.

Self-protection. The LS/CA detects unauthorised alterations to obfuscated operator policies stored in the system registry. It stalls operations while replacing the policies with securely obtained replacements. The LS/SAM performs the necessary traffic analyses to detect potential security threats and informs peer LS/SAMs, the overall network management system and/or the network administrator. In particular, the LS/SAM supports identification of malicious nodes that attempt denial of service attacks and blacklists them, warning the complementary access network management components.

4. The LS/ASAM Suite in Action

Ubiquitous data connectivity and communications management are optimised transparently across multiple network access technologies by dynamic coordination of the LS/ASAM components according to the specific situations. In particular, different combinations of their features enable a variety of deployment scenarios. In the following, two of the most significant ones are presented including a discussion of the distinctive characteristics in relation to relevant work.

4.1. QoS Enforcement in Heterogeneous Access Networks

The notion of guaranteed data transmission quality with enforcement mechanisms, in particular for emerging QoS sensitive multimedia applications, e.g., voice or video over IP, is a key issue especially in converged networks [10]. While traffic prioritisation is often not of paramount importance in core networks due to over-provisioning, QoS is an essential differentiator in limited-capacity wireless access networks for capacity and/or delay sensitive traffic such as voice or video over IP. While for cellular access technologies belonging to 2.5G, 3G and 3.5G, appropriate standards for QoS have been defined, few operators yet make widespread use of them. In addition, the WLAN world is supporting its technologies with specifications that directly account for QoS management.

In particular, when integrating different access network technologies, e.g., WLAN and UMTS, the quality of a connection may be degraded during vertical handover where (i) the connection needs to be re-established at the new access node, which is time consuming and during which no data can be transmitted, and (ii) if too many IP packets are lost, they must be retransmitted which can also be time consuming in the case of a large number of packets - again leading to service interruption.

Various approaches have been developed and proposed to address this problem. In [11], a reservation-based QoS model for integrated cellular and WLAN networks is defined and an adaptive mechanism to ensure end-to-end QoS is proposed. However, this model can only work by making the assumption that cellular/WLAN interworking is realised by relying upon a common and uniform reservation-based QoS architecture, which is not (yet) the case for most real network scenarios. Similarly, Song et al. [12] proposed an admission control mechanism for integrated voice and data services in cellular/WLAN networks. The main limitation of this approach though is that it does not account for video traffic.

To effectively provision QoS and optimise resource utilisation for a variety of possible heterogeneous network scenarios, the LS/ASAM Suite relies upon the dynamic combination of specific mechanisms both at the client side (i.e., seamless handover, session continuity and connection adaptation) and at the network side (i.e., congestion recovery and load-balancing) that are compliant with dominant industrial standards, e.g., mobile IP or SIP/IMS, when supported, or technology-independent, whenever possible.

Unlike legacy systems and hardware-based solutions, the LS/ASAM components accommodate high-level service and user needs and preferences (including QoS requirements) by implementing coordination mechanisms and resource allocation algorithms that hide low-level access technology dependent processes. This is achieved by deploying an agent-based middleware architecture that provides users with a common and higher level of abstraction, which makes low-level network access heterogeneity transparent.

On the client side, basic QoS in terms of service availability and continuity is enforced by the LS/CA through automatic and policy-driven vertical handover, i.e., all traffic is switched from one network interface, according to existing constraints and user policies. Moreover, by continuously monitoring network conditions and device status and properties, the LS/CA exerts QoS and context-aware resource management by selecting the most appropriate access technology to be used for the running applications/processes. In addition, when appropriate, as detailed in Section 3.3, the LS/CA can also trigger negotiation with one or more LS/SAMs for different connectivity conditions.

On the network side, the key mechanisms deployed by the LS/SAM to enforce QoS provisioning are load-balancing and congestion recovery. Load-balancing can be triggered by LS/SAMs in order to redistribute traffic across several access nodes according to various criteria, including:

- Current utilisation of resources at the access node, e.g., once the traffic overcomes a given threshold a certain portion of the supported connections might be handed over to neighbour LS/SAMs.
- QoS requirements of the running services, e.g., best-effort connections might be handed over to prioritise premium services for which charging might be based on service reliability guarantees (e.g., ≥95% non-disruption).
- Predictions of the network resources usage to minimise the probability of congesting an access node.

Analogously, whenever congestion occurs a specific part of the traffic at a given access node might be handed over to other LS/SAMs or selected existing connections (e.g., the non-premium ones) might even be dropped as appropriate. This enables relief of congestion and increases service resiliency and availability.

For example, assume a user that launches an IP-based TV program (e.g., a news channel) on a smart phone. During the launch of the selected application to render the video stream, the LS/CA determines the connectivity parameters (typically bandwidth and delay) for interruption-free high quality service provision. Because different access technologies offer different QoS assurances, the LS/CA might try to switch to a specific technology, e.g., UMTS, that better supports the QoS level needed for the video down-streaming. In addition, in the case of an UMTS connection, the LS/CA would set up a new Packet Data Protocol context requesting the UMTS QoS streaming class [13].

Figure 4 depicts the deployment model for this case. Each end user device is installed with an LS/CA component able to enforce QoS. The LS/CA must be aware of the different traffic categories available in each network access technology. During a vertical handover, the QoS class of the active network is mapped into an appropriate QoS class of the target network. There is one LS/SAM agent being deployed per access node, i.e., each LS/SAM agent is in charge of a specific access node and thus is up-to-date at all times regarding the status of that node. When planning load balancing and congestion recovery, the LS/SAM agent must be aware of the QoS classes supported by the different access technologies to minimise the risk of degraded service quality. This involves LS/SAM-to-LS/SAM coordination first to exchange information on current traffic load (or resource availability) and then to possibly take or hand over part of the communications/traffic[1].

4.2. Integration with an IMS/SIP Framework

IP Multimedia Subsystem, IMS, initially developed by 3GPP and 3GPP2 as an IP core network architecture for cellular/wireless-based access to Internet services, is now evolving into a standard that provides a common framework to create and offer next generation converged network services [14]. IMS builds on the Session

---

[1] Peer LS/SAMs coordination is not described in this paper because of some pending patenting issues.

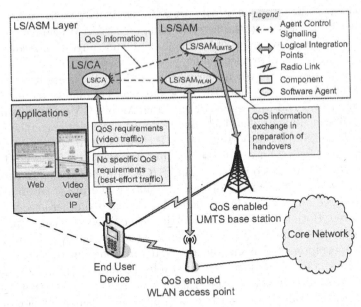

FIGURE 4. Deployment model of the LS/ASAM Suite for QoS enforcement.

Initiation Protocol, SIP, that is mainly responsible for delivering a session description to a user at her current location [15]. The key idea is to enable any kind of access (wireless or fixed) for any kind of media (including any combination of voice, text, image and/or video) supporting multiple devices and endpoints.

Because of the (at least initial) co-existence of IMS and non-IMS applications, the costs associated with moving to a full IMS-based network and the inherent complexity of IMS (and its several standards, interfaces and protocols) most service providers and/or operators are expected to migrate toward an IMS service framework iteratively.

One of the core issues to be addressed for successful adoption of IMS is the ability to face more aggressive bandwidth and latency demands, which implies increased QoS management and design capabilities on the bearer network [16]. In particular, IMS/SIP lacks traffic management capabilities and especially adaptive connectivity management and optimisation mechanisms that can be regarded as key components for delivering ubiquitous quality-sensitive multimedia services.

In this perspective, the LS/ASAM Suite complements an IMS-based framework by ensuring the quality of delivered services at the bearer network level through its adaptivity mechanisms, leaving IMS/SIP to cope with call control and service deployment issues. As depicted in Figure 5 the LS/CA component directly interacts with the SIP client installed on the end user device. In this way, the SIP client is able to obtain information on the quality of the connection which is helpful to determine, for instance, the appropriate codec to use, and to request the LS/CA component to ensure a certain quality level (in particular, when explicit QoS class enforcement is enabled). On the network side, an LS/SAM agent integrates with each access node and, by means of load balancing and congestion recovery, enables to provide a high level of service quality.

A simple use case is when one considers the collaboration between a SIP client and the LS/CA component to guarantee a level of quality required by a user to perform a video call (or, similarly, to watch Mobile TV). Upon launch of the SIP-based video calling application, the SIP client assesses the connection quality by means of the LS/CA component. The SIP client is aware of the quality requirements imposed by the video call service that are also variable according to the size and quality of the video picture. The LS/CA component can, in collaboration with the respective LS/SAMs, discover the quality offering at alternative access nodes and, based on that decide whether a handover to another access node needs to be triggered. Both end devices that participate in the video call must also agree on the codecs to be used for encoding and decoding the voice and video data. The LS/CA component delivers the necessary information to the SIP client to make its choice. Once the video call is established and running, it is the LS/CA agent's responsibility, in cooperation with the active LS/SAM agent, to preserve the quality of the connection and take appropriate measures if tolerance thresholds are violated. Depending on the mobility profile of the user, but also on the evolution of the network conditions, handoffs are unavoidable and thus need to be well planned and efficiently executed to minimise quality breaches.

The LS/CA does not affect the SIP call itself nor infringe any of the IMS/SIP standards. SIP is concerned with controlling the call execution while LS/ASAM takes care of connectivity. LS/ASAM is therefore complementary to IMS/SIP and benefits result even if only a small proportion of the entire network infrastructure (namely the access part) and end user devices are LS/ASAM empowered.

5. Experimental Analysis

In order to give a measure of the concrete benefits brought to a telecom operator by the adoption and deployment of a solution based on LS/ASAM, several experimental tests have been performed. This section first introduces the ASAM simulator, an instrument built for validating the basic concepts and evaluating various autonomic service access strategies on a set of simulated network settings representing real scenarios. One particular scenario is then selected to illustrate performance when different service access algorithms have been deployed in the user devices and in the access network. A set of preliminary experimental results are provided, obtained from the comparison of the discussed access strategies.

5.1. The ASAM Simulator

The ASAM simulator is an instrument built in Java for validating the ASAM concepts, in particular, how different autonomic access strategies deployed into LS/CA and LS/SAM modules should perform in real network access scenarios.

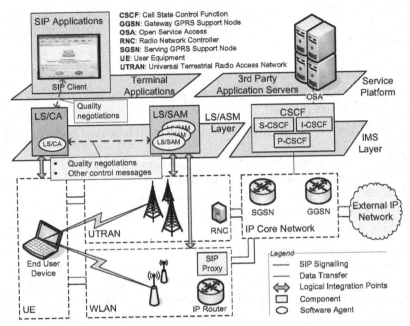

FIGURE 5. Deployment model of the LS/ASAM Suite when integrating with an IMS/SIP-based architecture.

In the real world people use their portable devices to request services in accordance with changes in location, activity and other requirements. Using radio communication they are able to connect to a network operator offering a heterogeneous infrastructure of different access node types (e.g., WLAN access nodes, GPRS/UMTS antennas, etc.) In the ASAM simulator, both user devices and access network components are modelled using software agents. Agents that simulate a user device can make use of the LS/CA where specific service access strategies are pre-loaded. In the same way, an agent representing an access network component can make use of the LS/SAM capabilities. The interaction between a device and an access node is then mapped through an exchange of FIPA-compliant messages. **Input Parameters.** Within the ASAM Simulator time is discrete and the simulations are based on the *quasi-static* condition. For this reason, input parameters related with the time are:

- Start time of the experiment.
- Duration of the experiment.
- Duration of a time step (e.g., 1 minute).

Furthermore, other parameters are required to describe the scenario:

- Locations represented in the experiment (e.g., train station, street, offices, etc.).

- Types of available network interfaces (e.g., UMTS, EDGE, etc.).
- Set of network services that are simulated in this experiment (e.g., phone call, VOD, email, etc.).
- Set of access nodes.
- Set of end user devices.

For each access node (e.g., WLAN access point, UMTS cell, etc.) the following input parameters are required:

- Type of network technology represented by this access node (e.g., UMTS cell).
- Nominal bandwidth of an access node measured in Bit/s.
- Maximum number of concurrent connections.
- Maximum bandwidth deployable on a single connection.
- Version of the LS/SAM the access node makes use of (e.g., none, *LS/SAM-BN*²).

Finally, for each user device to be simulated, the following input parameters are necessary:

- Location of the device at the beginning of the simulation (e.g., street).
- Set of network adapters installed in device (e.g., only GPRS).
- Amount of bandwidth that can be used at maximum given the network technology (e.g., 11 Mbit/s for WLAN).
- Set of service descriptors denoting the services that are available to the user who operates the device (e.g., a normal mobile phone can perform only calls).
- Version of the LS/CA the user device makes use of (e.g., none, *LS/CA-APF*²).
- A set of input parameters used to define a mathematical model which describes the behaviour of the end-user while using the device. This is defined in terms of movements among locations, usage rate and duration of services while being located in a given place. In particular, the following matrices must be provided:
    - The average time before a user changes her location, moving from one environment to another one.
    - The average time before a user issues a service demand while being located in a given space.
    - The duration of a started service while being located in a given space.
  The implemented mathematical model is based on the *Markovian property* that the probability of the occurrence of an event does not depend on the history of previous events. Based on this property, events like the starting of a service or the movement between locations are simulated with an occurrence rate equal to the inverse of the λ parameter of a *negative exponential* distribution. Furthermore, the duration of a started service is simulated using

---

²The suffix (APF in this case) determines the type of autonomic access strategies the component implements.

TABLE 1. Access nodes properties.

Type	Nominal Bandwidth	Max Bandwidth/Device	Max. Connections
WLAN	2000 Kbit/s	2000 Kbit/s	120
UMTS	1500 Kbit/s	300 Kbit/s	6
GPRS	400 Kbit/s	50 Kbit/s	10

the *Erlang-k* distribution. The expected average and standard deviation of the service duration are used to define the distribution.

In the ASAM simulator, each device has a user event generator that implements this mathematical model. The generated *user events* represent movements or service initiations with a stochastic duration. When an event occurs, the action is simulated (e.g., start a VOIP call in a road for 2 minutes). Each time an event is consumed, a new one is immediately generated. The generator also terminates elapsed services.

Output Variables. The ASAM simulator provides a set of output parameters that measure the performance of the LS/CA and LS/SAM strategies. The following list of output parameters includes only the subset of those used in Section 5.3:

- M_{ur}: The average used bandwidth of an access node in relation to its nominal bandwidth. High M_{ur} values encounter a high average utilisation of the access nodes which means that the infrastructure is more efficiently utilised.
- M_{sr}: The satisfaction rate of a demand is an indicator for the service quality that a user receives. Currently, this variable considers only the amount of bandwith consumed versus the amount requested.
- M_{fc}: The accumulated time span during which an end user device receives the bandwidth it requests and thus can deliver full service quality to the user. Values are normalised in the range [0...1].
- $M_{d.vho}$: The average occurrence rate of vertical handoffs in a time step when triggered by a user device.
- $M_{n.vho}$: The average occurrence rate of vertical handoffs in a time step when triggered by an access node.

5.2. Simulation Setup

This section presents preliminary laboratory experiments conducted to validate the ASAM concepts through the use of the ASAM simulator. The presented simulation evaluates what might happen in a normal working day during which a large number of people arrive at a train station before dispersing to their places of work where they spend most of their day.

Figure 6 illustrates the simulated access network topology where different access nodes (UMTS/GPRS cells and WLAN access points) cover different locations (a train station, three roads and two offices). The access nodes exhibit the characteristics presented in Table 1. The reported values are similar to the characteristics offered by typical network components deployed in most access networks.

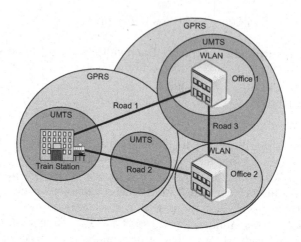

FIGURE 6. Access network topology used in the presented simulation.

TABLE 2. Average movement rate exhibited by the users.

To From	Train Station	Road 1	Road 2	Road 3	Office 1	Office 2
Train Station	-	5 mins	5 mins	-	-	-
Road 1	10 mins	-	-	-	5 mins	-
Road 2	10 mins	-	-	-	-	5 mins
Road 3	-	-	-	-	10 mins	10 mins
Office 1	-	5 hours	-	4 hours	-	-
Office 2	-	-	5 hours	4 hours	-	-

In this simulation, 40 users, starting from the train station, move around this scenario with their devices consuming network services. Their devices are able to handle communication with all the available network technologies (WLAN, UMTS and GPRS).

Table 2 describes the average movement rates exhibited by the users. It is important to notice that these rates are unidirectional, that is, the frequency of moving from one location to another is not necessarily the same of the reverse direction between the same locations.

The information about how services are used is described in Table 3. In particular, the table presents the average occurrence rate and the duration of a service, given the user's location. Moreover, the service duration is described in terms of the average time and its standard deviation. Additionally, the bandwidth consumed by each type of service is defined in the table column headers. Values in Table 2 and Table 3 do not represent a specific case but we believe these quantities are sufficient to analyse how the selected access strategies behave.

TABLE 3. Occurrence rate and duration of services started in specific locations.

	eMail 80Kbit/s	VOIP 128Kbit/s	Internet 240Kbit/s	VOD 1Mbit/s
Train Station	40 mins 5 mins ± 1	2 hours 2 mins ± 1	3 mins 5 mins ± 1	-
Road 1	1 hour 5 mins ± 1	2 hours 2 mins ± 1		
Road 2	1 hour 5 mins ± 1	2 hours 2 mins ± 1		
Road 3	1 hour 5 mins ± 1	2 hours 2 mins ± 1		
Office 1	20 mins 5 mins ± 1	1 hour 10 mins ± 1	1 hour 15 mins ± 2	4 hours 30 mins ± 5
Office 2	20 mins 5 mins ± 1	1 hour 10 mins ± 1	1 hour 15 mins ± 2	4 hours 30 mins ± 5

The described scenario was simulated for 24 hours with a time step of 1 minute. Given the non-deterministic property of the experiments, 10 simulation runs of the same scenario have been performed.

Network access strategies. In this experiment we consider two network access strategies: one implemented by the LS/CA and one implemented by the LS/SAM.

The access strategy implemented in the LS/CA equipped user devices, named *Adapter Priority Function* (APF), assigns a dynamic priority function at each adapter of the user device. This function takes as input measured and expected parameters values that are: battery power, time since last handover, used bandwidth, end-to-end delay, adapter statistics, adapter cost and creates a weighted linear combination of a set of sub-functions built on the listed parameters. If the adapter with the highest function value is different from the currently used, a handover is triggered. We name an LS/CA that implements the APF access strategy as *LS/CA-APF*.

The access strategy implemented in the LS/SAM equipped access nodes, named *Balance* (BN), tries to keep the quality of the required services high balancing the load among the access nodes available in the user device's neighbourhood. Whenever an established connection obtains less bandwidth than the requested one, the access node using a Contract-Net protocol asks to other nodes how much bandwidth they could offer to that connection. The candidate access node should be able to satisfy the requested bandwidth and minimise the gap between the bandwidth demand and the bandwidth offered. If there are no access nodes that offer more than the requested bandwidth, the connection is assigned to that access node with the highest bandwidth offered. If no proposals are better than what the

TABLE 4. Comparison of the results obtained simulating four different network access configurations.

	Without LS/ASAM	Only LS/CA-APF	Only LS/SAM-BN	With LS/ASAM
M_{ur}	0.2821 ± 0.0062	0.2897 ± 0.0265	0.4048 ± 0.0066	0.4199 ± 0.005
M_{sr}	0.5433 ± 0.0070	0.5839 ± 0.0637	0.7382 ± 0.0137	0.7621 ± 0.0138
M_{fc}	0.1153 ± 0.0088	0.1540 ± 0.0752	0.2216 ± 0.0156	0.2583 ± 0.0260
$M_{n.vho}$	0	0	0.1685 ± 0.0212	0.1478 ± 0.0064
$M_{d.vho}$	0.0169 ± 0.0008	0.0066 ± 0.0006	0.0095 ± 0.0010	0.0015 ± 0.0001

current access node offers, no handover is performed. We name an LS/SAM that implements the BN access strategy as *LS/SAM-BN*.

If both the LS/CA and the LS/SAM are deployed in the network, that is, the whole LS/ASAM system is in use, a mechanism to avoid conflicts between provider and user strategies is adopted.

In order to understand the benefits provided by LS/ASAM, the scenario where LS/ASAM is not present was also simulated. In this case, access nodes do not exhibit any access logic and the devices select the preferred access node based on the highest nominal bandwidth a network technology provides (e.g., WLAN, UMTS, GPRS).

5.3. Results

Table 4 presents the results obtained simulating four different network access configurations: the case without the LS/ASAM system, the case with only LS/CA-APF components, the case with only the LS/SAM-BNs, and the last case where the whole LS/ASAM system is deployed.

The results show that if only LS/CA components are deployed in the network (third column), they are able to improve all the evaluated metrics when compared with the case where no LS/ASAM components are in place. For example, LS/CAs generate 33% more time where users receive the requested bandwidth even with a lower number of vertical handovers. .

Results are even better if we compare the case without LS/ASAM to the case where only LS/SAMs are deployed. In this case, for example, the user satisfactory metric, M_{fc}, improves by 100% and the usage rate metric, M_{ur}, improves by 42%. From this we can conclude that an operator may be able to mitigate the need for extensions to network infrastructure in lieu of deploying some software intelligence into existing infrastructure.

Moreover, we can observe from these results that adding intelligence in the access network brings about greater benefits than adding intelligence to user devices. This is because the network has a broader and real knowledge of the current infrastructure status than a user device that also bases its decisions on estimated values.

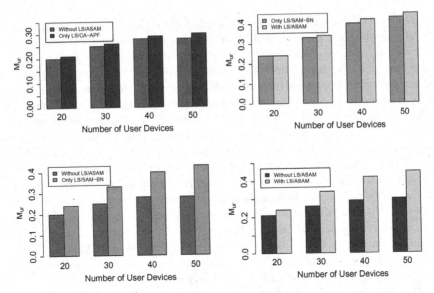

FIGURE 7. The benefits provided by LS/ASAM (or some of its components) against an increasing number of network users.

Furthermore, we can notice that the combined use of LS/CA and LS/SAM components (the whole LS/ASAM system) generates yet greater benefits than those ones generated by one of the two components in isolated use. We believe that further improvements of the evaluated metrics will be obtained with as yet to be reported work regarding the simulation of collaborative strategies between LS/CA and LS/SAM as presented in Section 3.3.

Finally, Figure 7 shows the benefits provided by LS/ASAM (or some of its components) with the increasing of the network users when analysing the M_{ur} metric, that is, usage of the network. Considering for example the histogram in the bottom-right of the figure, where a system with LS/ASAM deployed is compared to one without, it is notable that as the volume of users increases, so does the gain achieved with LS/ASAM.

To measure the real significance of the obtained results the Wilcoxon Paired Rank Sum Test was applied. This test stated with a confidence level higher than 95% that the improvements generated by the LS/ASAM system are statistically significant.

6. Discussion and Conclusions

The LS/ASAM Suite is a distributed and resilient system that exhibits high adaptivity to its network environment. This has been achieved by properly combining multi-agent systems concepts and technology with powerful resource allocation algorithms and reasoning strategies.

The central idea is that loosely-coupled distributed management functions and control methods can be well-modelled and implemented by making use of automated, goal-driven and proactive software entities. These lightweight components are able to operate on resource-scarce devices and support asynchronous communication with intermittent network connections. Moreover, according to the results of proactive monitoring information received from the environment within which they are embedded, the LS/ASAM components directly assist with autonomic management of network resources. They are able to configure themselves and dynamically optimise their operations according to the way their environment changes and in-line with operator and client user policies. They thus assist with the speed-up and automation of simple, tedious and repetitive service management tasks currently performed most commonly by human operators. The ultimate result of this is potentially substantial cost savings to the operator. In particular, by hiding low-level networking aspects that, especially in converged network scenarios, can continuously change due to end users mobility, the LS/ASAM middleware provides transparent service access in heterogeneous networks and becomes an essential complement to (bearer unaware) service delivery platforms.

However, to achieve the potential of autonomic management systems in todays' networks is not a straightforward task. Migrating intelligence and complex management functions toward the edge of the network reduces the degree of manual intervention needed, but increases somehow the complexity of the management system itself. The network has indeed to be adaptable, but at the same time stable and controllable. Therefore, populating the networking environment with autonomic software components requires some additional configuration and monitoring capabilities. In this sense, middleware technologies for highly dynamic and heterogeneous networks must become able to monitor and control the middleware itself by integrating with traditional, relatively static infrastructures often populated by legacy solutions and adapting to different operating systems and connection technologies. This is a challenging task that still requires additional investigation.

The system described in this paper has been implemented as a fully-functional prototype with an accompanying scenario simulator for experimental evaluation. The results presented in this paper are rather preliminary in that we cannot yet report on the fully mediated solution, but they are nevertheless extremely encouraging. In particular, the demonstrable performance improvement with the complete LS/ASAM suite in operation, as shown in Figure 7, is a quite significant result.

Our ongoing and future work includes more refined and extensive characterisation of LS/ASAM performance, especially on the network side, when adopting different user and operators' policies, network allocation strategies and algorithms. While the LS/CA has been already successfully deployed in a variety of real-world scenarios, the adoption of the LS/SAM requires some additional work given the wide assortment of existing and upcoming service and network management architectures. In particular, by simulating and analysing the LS/ASAM Suite performance in a variety of networking scenarios and consequently refining the behaviour of the various system components, we expect to better characterise and consequently improve performance and scalability. In addition, by means of selected testbed demonstrations and experiments we are assessing the feasibility and complexity of integrating LS/ASAM entities in specific service delivery frameworks including IMS.

Acknowledgment

Many thanks to the colleagues at Whitestein Technologies who contributed significantly toward this work, in particular Thomas Lozza, Martin Stangel, Oliver Hoeffleur, Oliver Carl and the LS/CA team.

References

[1] J. Strassner, *Autonomic Networking - Theory and Practice.* Proceedings 9th IFIP/IEEE International Symposium on Integrated Network Management, Nice, France, (2005).

[2] N. R. Jennings, *Agent-Oriented Software Engineering.* Lecture Notes in Computer Science, **1647**, 1999, 1–7.

[3] R. Ferrus, A. Gelonch, O. Sallent, J. Perez-Romero, *Vertical Handover Support in Coordinated Heterogeneous Radio Access Networks.* Proceedings 14th IST Mobile and Wireless Communications Summit, Dresden, Germany, 2005.

[4] C. Perkins, *RFC 3220: IP Mobility Support for IPv4.* 2002.

[5] S. Kent, R. Atkinson, *RFC 2401: Security Architecture for the Internet Protocol.* 1998.

[6] M. Yokoo, K. Hirayama, *Algorithms for Distributed Constraint Satisfaction: A Review.* Autonomous Agents and Multi-Agent Systems, **3:2**, 2000, 185–207.

[7] Foundation for Intelligent Physical Agents, *FIPA Iterated Contract Net Interaction Protocol Specification.* 2001.

[8] J. Hofbauer, W. H. Sandholm, *On the global convergence of stochastic fictitious play.* Econometrica, **70**, 2002, 2265–94.

[9] J. S. Shamma, G. Arslan, *Unified convergence proofs of continuous-time fictitious play.* IEEE Transactions on Automatic Control, **49:7**, 2004, 1137–42.

[10] K. Ahmavaara, H. Haverinen, R. Pichna, *Integration of wireless LAN and 3G wireless - Interworking Architecture between 3GPP and WLAN systems.* IEEE Communications Magazine, **41:11**, 2003, 74–81.

[11] X. G. Wang, G. Min, J. E. Mellor, K. Al-Begain, L. Guan, *An adaptive QoS framework for integrated cellular and WLAN networks.* Computer Networks, **47:2**, 2005, 167–183.

[12] W. Song, H. Jiang, W. Zhuang, X. Shen, *Resource management for QoS support in cellular/WLAN interworking.* IEEE Network, **19:5**, 2005, 12–18.

[13] 3GPP, *TS 23.107 V7.4.0: Quality of Service (QoS) concept and architecture.* 2006.

[14] M. Cuevas, *Admission control and resource reservation for session-based applications in next generation networks.* BT Technology Journal, **23:2**, 2005, 130–145.

[15] M. Kolbehdari, D. Lizotte, G. Shires, S. Trevor, *Session Initiation Protocol (SIP) Evolution in Converged Communications.* Intel Technology Journal, **10:1**, 2006.

[16] ABIresearch, *IP Multimedia Subsystem Industry Survey Results.* 2005.

Monique Calisti, Roberto Ghizzioli and Dominic Greenwood
Whitestein Technologies AG
Pestalozzistrasse 24
CH-8032, Zürich,
Switzerland
e-mail: {mca,rgh,dgr}@whitestein.com

Whitestein Series in Software Agent Technologies, 127–148
© 2007 Birkhäuser Verlag Basel/Switzerland

Cross-layer Optimisations for Autonomic Networks

Mohammad Abdur Razzaque, Simon Dobson and Paddy Nixon

Abstract. Autonomic communication aims to build more reflective systems with properties like self-healing, self-organisation, self-optimisation and so forth – the so-called "self-*" properties. To attain this within the existing strictly-layered approaches to network software may be possible to certain extent, but will not leverage all the possible optimisations, and we suggest that cross-layer architectures are better-suited to achieving the self-* properties. This paper explores the possibilities of cross-layering approaches in autonomic networks, reviews and compares the different cross-layer approaches to network architecture, observing that most current approaches depend purely on local information and provide only poor and inaccurate information gathering at the network level.

1. Introduction

The current convergence of networked infrastructures and services has changed the traditional view of the network from the simple wired interconnection of a few manually-administered homogeneous nodes to a complex infrastructure encompassing a multitude of different technologies (wired/wireless, mobile/fixed, static/*ad hoc*), heterogeneous nodes (in terms of size, capabilities, power and resources constraints as well as platform) and diverse services (end-to-end, real-time, QoS). This situation poses a challenge for the research community to engineer systems and architectures that will increase the QoS and robustness of the current and future Internet whilst alleviating the management cost and operational complexity. On the other hand the external simplification of TCP/IP has not unfortunately been matched by a corresponding simplification in the construction, management and extension of the network from a provider's perspective. Adding a new network segment, a new protocol, a new kind of element or support for a

This work is partially supported by Science Foundation Ireland under grant number 04/RPI/1544, "Secure and Predictable Pervasive Computing."

new user- or system-level application have become fraught exercises in managing the complexity of interactions between elements. This in turn both reduces innovation in networks and network-centric services and can directly affect the economic viability of products and services that rely directly on IT and communications agility.

Existing network paradigms deal poorly with multilevel tension between complexity and simplicity, diversity and ubiquity. Traditional networks have been constructed and coordinated centrally according to a single plan and can consequently be architected using a homogeneous population of components with common technical standards and management goals. By contrast, next-generation networks are expected to grow more chaotically with no centrally-mandated goals or levels of service, no universally agreed upon protocols or other technical standards, and no *a priori* knowledge of the topology or component population. Specifically, the next-generation network must be radically distributed and decentralised, self-describing, self-organising, self-managing, self-configuring and self-optimising, providing a seamless communications infrastructure composed of multiple technologies and able to leverage local information and decisions without sacrificing global performance, robustness and trustworthiness.

The development of self-managing, self-configuring and self-regulating network and communications infrastructures – collectively referred to as autonomic communications – is an area of considerable research and industrial interest. By analogy to the human autonomic nervous system, which regulates homeostatic functions without conscious intelligent control, autonomic communications seeks to simplify the management of complex communications structures and to reduce the need for manual intervention and management. It draws on a number of existing disciplines including protocol design, network management, artificial intelligence, pervasive computing, control theory, game theory, semantics, biology, context-aware systems, sensor networks, trust and security. The distinguishing feature is the fusion of techniques from these fields in pursuit of a goal of simplified systems deployment and management.

The tremendous proliferation of today's Internet is mainly due to architecture based on the TCP/IP layering. Even with its (strict layering) success, however, this trend may be criticised as providing too narrow an interpretation of the information that can usefully be made use of at a particular layer of abstraction in a complex software system. By reducing the information available to a minimum in the interests of simplicity, it is possible that some opportunities for optimisation are lost. In particular, given the rise of autonomic communications, we would contend that contextual information of vital use in adapting the behaviour of a network to its use and environment is being neglected, and that this acts as a brake on the creation of self-managing, self-adaptive autonomic communication systems. All of these above issues have posted a fundamental question: is layering still an adequate foundation for future Internet and network architecture?

Would a completely non-hierarchical, non-layered, "flat" approach be a better solution for the next-generation Internet? Such completely non-layered designs can lead to various negative consequences. In this situation an alternative solution to non-layered approach is the modification of strict layering – cross-layering – which increases the amount of information available to a network sub-system about the content it is carrying and the context in which it is operating. It is clear from the recent initiatives in autonomic computing and autonomic communications [6, 10] that there is a need to make future networks self-behaving, in the sense that they work in an optimal way with "endogenous" management and control, and with minimum human perception and intervention.

A number of proposals for cross-layer designs and their corresponding architectures have been published in the literature. Alongside the categorisations mentioned in [16], all of the existing cross-layer design architectures can be classified according to how they are getting the information for the optimisations: (i) architectures based on local information only (from the node and its different layers); and (ii) architectures based on both local and non-local information (from the node, its different layers and from neighbours). Most existing architectures (including GRACE [18], WIDENS [11], MobileMan [3]) are based on a local view of the state of, and constraints on, the network: only CrossTalk [20] is based on a global view (even partially). On the other hand, POEM [8] is one of the very few architectures considering self-optimisation that could be helpful for autonomic communication. For system wide optimisation we need network-wide non-local information alongside the local view. Considering this, [15] proposes an architecture with local and as well network-wide information exchange to show that cross-layer optimisation can provide autonomic features for communications networks.

This article explores the possibilities of cross-layering approaches in autonomic networks, reviews the cross–layer approach to network architecture and compares the different cross-layering architectures, observing that most current approaches depend purely on local information and provide only poor and inaccurate information gathering at the network level.

The rest of this article is structured as follows. Section 2 describes autonomic and traditional networking and their limitations, including a brief discussion of the business forces driving its evolution. Section 3 presents the alternative architectures in networking and discusses the possible use of cross-layer architecture in autonomic communications. Section 4 concludes with some future directions for the application of cross-layer design to autonomic networks.

2. Autonomic versus traditional networking

Developers of large-scale applications have experienced increased complexity in their software systems due to the relentless integration of new services. This affects not only software developers but also the service providers who have to manage these software systems.

Traditional businesses have generally treated IT systems and networks as cost centres, used in the back-office delivery of front-office services back-office delivery of front-office products and services. However, the increasing penetration of Internet-delivered services has opened companies' networks to their customers and suppliers, allowing tighter integration and reduced complexity of interactions. Furthermore many companies are now offering services on-line that streamline their more traditional offerings – or indeed that have no off-line analogue. This raises the possibility of networks becoming a profit centre for businesses able to drive value from them.

What are the business requirements generated by such a transition? Firstly, the business opportunities that can be addressed directly by networking tend to be short-lived, implying that it must be possible to develop and deploy new features very quickly. This in turn implies that adding new features to existing systems must be straightforward, and must not compromise other services being provided simultaneously.

Secondly, opportunities are often targeted at only specific customer segments. Driving maximum value from such a service implies that the customers of interest can be reached by networked applications, and that all possible customers are in fact targeted. This places a "forward" requirement on services to operate as broadly as possible; however, it also introduces a "backward" requirement in that customers may profile themselves through their uses of other services, and a flexible network infrastructure may provide an accurate means to customer segmentation.

Broad operation means that networks must operate to deliver services regardless of the changes that occur around them. This leads to the third business driver, the need for uniform quality of service across delivery channels. An inadequate delivery channel is often worse than no channel at all: it can cause severe brand damage that can contaminate other offerings. Combined with rapid change, it is unlikely that a human operator will be able to monitor and optimise all possible access channels, so the responsibility must perforce be accepted by the network itself: using technology to manage the functioning of technology.

Such self-management is extremely complex, at two levels. At the technological level it involves monitoring a network, identifying problems and formulating strategies to address them in a timely and correct manner. However, above and beyond this the strategies selected must be correct from a business perspective as well as from a purely technical one: it does not good to (for example) reduce a network's congestion by reducing the quality of a video stream when that video stream is the most valuable content being delivered by the network. Such prioritisation between approaches implies that networks must take account of factors arising from several different conceptual layers within a system.

It is important to remember that development and deployment (capex) form only a minor part of the overall cost of ownership in which operational costs and maintenance (opex) dominate. Deploying a system rapidly may provide access to a market opportunity, but that opportunity, and the revenues deriving from it, may

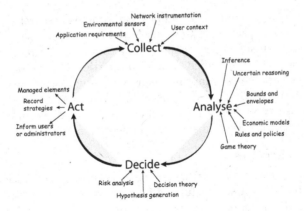

FIGURE 1. Autonomic control loop (from [6])

then be squandered if system maintenance, provisioning and dimensioning are not considered from the outset.

Existing network paradigms deal poorly with this multilevel tension between complexity and simplicity, diversity and ubiquity. Moreover next-generation networks are expected to grow more chaotically with no centrally-mandated goals or levels of service, no universally agreed upon protocols or other technical standards, and no *a priori* knowledge of the topology or component population.

To tackle these issues next-generation network has to be fundamentally distributed and decentralised, self-organising, self-managing, self-configuring and self-optimising, providing a seamless communications infrastructure composed of multiple technologies and able to leverage local information and decisions without sacrificing global performance, robustness and trustworthiness.

In 2001, IBM launched the autonomic computing initiative [10] as a counter to the increasing complexity of software systems, aiming specifically at reducing the total cost of ownership of complex systems. The approach has since been broadened to include autonomic communications, characterised as service-driven, situated, autonomously controlled, self-organised, distributed, technology independent and scalable communications and applications platforms [6].

An autonomic system is a system that operates and serves its purpose by managing its own self without external intervention even in case of environmental changes. Autonomic systems form a feedback loop (Figure 1): the system collects information from a variety of sources including traditional network sensors and reporting streams but also including higher-level device and user context. These are analysed to construct a model of the evolving situation faced by the network and its services, with this model used as a basis for adaptation decisions. These decisions are actuated through the network and will potentially be reported to users or administrators. The impact of the decisions can then be collected to inform the next control cycle.

The vision of autonomic communication systems research is that of a networked world in which networks and associated devices and services will be able to work in a totally unsupervised manner, able to self-configure, self-monitor, self-adapt and self-heal – the so called "self-*" properties. On the one hand, this will deliver networks capable of adapting their behaviours dynamically to meet the changing specific needs of individual users; on the other, it will dramatically decrease the complexity and associated costs currently involved in the effective and reliable deployment of networks and communication services.

2.1. Layered architectures

The layered architecture for networking, on which the current Internet architecture is loosely based, is a successful example of architectural approach. The OSI reference and TCP/IP models are the classic examples of strict layer architecture [17].

The OSI Reference Model. The Open System Interconnection (OSI) Model was the first step towards an architecture for interconnecting various systems. It was aimed to break down a complicated system into manageable modular components and systematically specifying the interactions among these components. The main contribution of OSI model is the concept of layering that separates services, interfaces and protocols. Service is the function/semantic of a layer offered to an upper layer; interface tells how a layer can be accessed; and protocol is the internal implementation of a service. Layering offers the abstraction needed to ease complexity while understanding the complete system. Modularity in the form of stacked protocol layers can greatly accelerate the design, implementation, operation and management of a system through parallel efforts on individual modules. Interoperability is ensured by the standardised interfaces between the layers. Individual modules can be upgraded separately without forcing a system redesign, and the new modules are sure to interoperate with the others.

Although a major conceptual success the OSI model failed as an implementation strategy for a number of reasons [17]: the complexity of the model itself, the difficulty in implementation, inefficiency in operation, duplicated functions across the layers and some seldom used layers buried the model. Too strict layering hampered the performance, and consequently the development, of advanced network stacks.

The TCP/IP Model. The extraordinary proliferation of today's Internet is mainly due to its architecture based around the TCP/IP Reference Model. TCP/IP was from the very beginning protocol-centric. The TCP/IP model, as a re-description of the protocols, is more concise, containing only four layers. There is actually no strict layering like OSI, and applications are free to bypass the transport layer and go directly to the underlying IP protocol or even the network interface, with the prerequisite that interactions must be conducted through the controlled interface provided by the protocol headers. The flexibility is further enhanced in that any protocol can be inserted into the architecture, as long as the protocol specification and running code are provided and pass the standardisation procedure.

However, the model is far from perfect. Noticeable, there were originally no clear distinctions between services, interfaces, and protocols, as everything is expressed in terms of protocols. The object-oriented good points possessed by the OSI model like the separation of specification and implementation, are missing. To make things worse, the network interface layer is actually an interface rather than a layer in normal sense, and the boundary between the data link layer and the physical layer vanishes. Although design and implementation of TCP and IP were well done, the other supporting protocols are more or less *ad hoc* [17].

These strictly layered approaches are quite successful in providing best-effort service through the Internet, which is mostly dominated by wired network. However, TCP in particular is a single-resource-based protocol (bandwidth) which is optimised for a wired network, not for wireless. Present and next-generation communications are, however, hugely dominated by wireless and mobile computing, and require more flexible approaches to be taken. Specifically, next-generation networks must deal with both business trade-offs and technical limitations, restrictions on capacity and the need to structure that capacity at a higher level. To make the networking more "meaningful" in this sense we need to identify the limitations of the existing strict layering approach.

2.1.1. The Limitations of Strict Layering.

One obvious shortcoming of the two classical network reference models is the lack of information sharing between the protocol layers. Layers force narrow interfaces which can lead to "semantic squeezing" [5] where two phenomena that are distinct at one layer become indistinguishable to lower layers and so cannot benefit from different transport strategies. This hampers optimal performance of the networks, since shared layer information is the prerequisite for many forms of performance optimisation.

OSI and TCP/IP support a bottom-up approach driven by physical and network constraints, which makes it hard to capture and respond to top-down user demands or requirements. Introducing a single co-located layer for various adaptation tasks would be too complex and heavyweight, as well as being inadequate: QoS adaptation requires the participation of all layers [9]. A co-operative solution involving coordinating the individual adaptations of multiple layers would lead to a more flexible approach, although introducing the potential for feature interaction and instability.

The assumptions in the wired IP stack are inadequate for wireless networking, and TCP is known to suffer from performance degradation in mobile wireless environments. This is because such environments are prone to packet losses due to high bit error rates and mobility-induced disconnections. TCP interprets packet losses as an indication of congestion and (inappropriately) invokes congestion control mechanisms, which leads to degraded performance.

Wireless networks offer several possibilities for opportunistic communication that cannot be exploited sufficiently in a strictly layered design. Furthermore, the wireless medium offers some new modalities of communication the layered architectures do not accommodate, for example making the physical layer capable of receiving multiple packets at the same time [9].

Context awareness is a key issue in autonomic computing and networks. Context may be defined as "any information that can be used to characterise the situation of entities" [4] (whether a person, place or object) that are considered relevant to the interaction between a user and an application, including the user and the application themselves. The autonomic control loop of Figure 1 shows the usefulness of context in an autonomic system. This context can be use to improve performance, adaptability, user satisfactions and so forth. Context-awareness and adaptation can help significantly in network management.

The strictly layered approaches allow interaction only between adjacent layers and this restricts the possibility of context-awareness within different layers and user. Any approach which supports interactions between non-adjacent layers might be helpful in overcoming these obstacles.

We must however keep in mind that the concept of a non-layered protocol architecture has immediate implications. Layering provides modularity, a structure and ordering for the processing of meta-data, and encapsulation. Modularity, with its opportunity for information hiding and independence, is an indispensable tool for system design. Any alternative proposal must provide modularity, but also adequately address the other aspects of layering.

2.2. Providing self-* behaviours in layered systems

Autonomic networking's self-* properties require some node-wide local knowledge and non-local (possibly network-wide) knowledge about the behaviour of network elements [14]. Local information includes some sorts of user preferences and requirements, protocol layer information and node capacity; non-local information includes neighbouring nodes' local states, processing capacity, processor utilisation, traffic patterns and remaining power. Generally a network-wide view is a summation of summarised information regarding all the nodes in the network. To generate a network-wide view, the node-local views are generated first. Generation of local view requires different layers in protocol stack have to interact with each other, which may include non-adjacent layer interactions – which may not be supported in a strictly layered architecture. For example, self-optimisation in an autonomic network is stack-wide instead of layer-wide. If we consider self-organisation, this also requires non-adjacent layer interactions as well as the network-wide information. So to attain such a self-* system within existing strictly-layered approaches may be possible, but will not (we claim) easily leverage all the possible optimisations. Systems that support non-adjacent layer interactions may well offer a better prospect, if they can be structured to avoid chaotic interactions that will be impossible to maintain in the long term.

3. Alternative protocol architectures

To overcome the limitations of strict layering, researchers are considering alternative architectures like de-layered architectures, protocol virtualisation and – perhaps the most promising – cross-layer architectures.

3.1. De-layered architectures

One may think completely non-hierarchical non-layered and flat architectures could be a better solution for next generation Internet and meaningful networks. This is very much a "revolutionary" approach, which is free of existing layered concept and provides new and more philosophical definitions of functional entities and their interactions. It is concentrated more on the performance (in the broad sense) and does not compromise to maintain compatibility, which weakens its universal dimension and makes its deployment significantly more risky and expensive.

Role-base architecture. Role-based architecture or RBA [7] is an example of this approach which presents a new way to organise the protocols, in heaps not in stacks as done by layering. However, in addition to the compatibility issues this de-layered design can lead to various negative consequences. The layered architecture and controlled interaction enable designers of protocols at a particular layer to work without worrying about the rest of the stack. Once the layers are broken through no-layer interactions, this luxury is no longer available to the designer. There may be interdependency between some parts of non-layered approach and unbridled non-layer interactions can create loops, and from control theory's point of view, become a hazard to the stability of the system. Finally, completely removing layers will create an isolated "island" of connectivity which cannot communicate with the existing huge legacy of layered networks and Internet.

3.2. Protocol virtualisation

Distributed systems have often struggled against problems of scalability in their physical provisioning. A system that is scalable in theory must still be able to deploy enough computational and communication resources to handle increasing loads – and conversely scale its resource usage as demand decreases. The problem has been that increasing resource availability involves deploying machines, which can only be carried out on human (rather than machine) timescales and so cannot adapt rapidly. Moreover an individual machine may not be the appropriate unit of resourcing for many services, either too large or too small.

A recent approach to this issue is to "virtualise" the computing platform. A collection of computer servers is deployed, with each server emulating a number of smaller machines. Each virtual machine runs a complete computing environment – typically Linux or some other Unix variant – which is then mapped onto some portion of the host server's physical resources.

The advantages of this approach are five-fold:

- Each virtual machine can be dedicated to a single service

- A single host machine might support virtual hosts offering different operating systems
- Resources can be allocated to services in smaller units, relating to virtual rather than real (and perhaps excessively capable) machines
- New virtual machines may be allocated (and re-allocated) on-line up to limit of the host machines, allowing more dynamic provisioning as seen by the services themselves
- Each individual virtual machine is isolated from the others, meaning that crashes or virus infestations are less likely to propagate

The canonical examples of machine virtualisation are commercial systems such as VMWare and Xen, and open-source solutions such as user-mode Linux.

These techniques can be extended to protocols. A service uses standard application-level protocols (for example TCP/IP sockets) which are then implemented in non-standard ways. Rather than mapping a TCP connection to a standard stream of IP packets, the connection is instead carried on some other protocol transparently of the service. The service protocol is "virtual" in the sense that a single service-level protocol may be carried over several different underlying wire protocols.

The advantages of this approach are, again, flexibility and dynamism. Especially for wireless networks, the traditional TCP architecture can be very expensive, since it interprets packet loss as congestion rather than – as is often the case for wireless systems – as occlusion or interference. However, in a heterogeneous network there may be little choice but to use TCP as the service protocol of choice: the service may not be able to deploy another, more wireless-friendly protocol. Virtualising the protocol means that the server or network operating system may dynamically select a carrier protocol for each connection rather than globally for the network, or even that the same TCP stream can be carried on other protocols at different points along its journey.

The costs and performance of such systems are likely to be opaque; since it is difficult a priori to decide how a particular stream will be carried. Significantly more work is required to determine the best way to perform protocol selection and conditioning. We believe nonetheless that such virtualisation techniques are potentially and extremely valuable contributions to autonomic communications, and may work extremely well with cross-layer approaches that allow flexible but structured access to context.

3.3. Cross-layer approaches

The traditional layered design of network protocols is insufficiently flexible to cope with the dynamics of wireless-dominated next-generation communications. Recent studies [16] show that careful exploitation of some protocol interactions that cross the normal layer boundaries can lead to more efficient performance of different wireless scenarios [13, 14, 15].

3.3.1. What is Cross-Layer Networking? Cross-layer design breaks away from the traditional network design, where each layer of the protocol stack operates independently and information is exchanged only between adjacent layers via narrow interfaces. Information in cross-layer architecture is exchanged between non-adjacent layers of the protocol stack, typically using a broader and more open data format, and end-to-end performance is optimised by adapting to this information at each protocol layer. Cross-layering is therefore not the simple replacement of a layered architecture, nor is it the simple combination of layered functionality: instead, cross-layering attempts to share information amongst different layers, which can be used as input for algorithms, for decision processes and adaptations.

The motivations for cross-layer design [13] are very clear from the challenges posed by strict layering. Three main reasons for cross-layering may be summarised as the unique problems created by wireless links, the possibility of opportunistic communication on wireless links and the new modalities of communication offered by the wireless medium.

3.3.2. Existing Cross-Layer Architectures. Cross layer could be an attractive solution to improve the performance of wireless networks, to support autonomics in networking, etc. However, as cross layer research in the very early stage, research is ongoing to search out a generic cross layer infrastructure or architecture. And it is clear from [13] the importance of a good and sound architecture for proliferation cross-layer in wireless as well other communications era especially in autonomic communication networks. Within its short history of research, a number of proposals for cross layer designs and their corresponding architectures have been published in the literature. Most of these cross layer design proposals are based on one of the following basic categories [16]:

- Creation of new interfaces: In this type new interfaces between the layers are created and the new interfaces are used for information sharing between the layers at run-time.
- Merging adjacent layers: In this category two or more adjacent layers design together such that the service provided by the new super-layer is the union of the services provided by the constituent layers. This does not require any new interfaces to be created in the stack.
- Design coupling without interfaces: This design category involves coupling two or more layers at design time without creating any extra interfaces for information sharing at run-time.
- Vertical calibration across layers: This design category utilises vertical calibration across different layer's parameters to obtain optimum performances.

Moreover all of the existing cross layer design architectures follows one of the following categories to implement the cross layering interaction:

- Direct communication between layers: This type of interaction allows different layers to expose their information or variables during run-time to optimise overall performance. CLASS [19] is one of the examples of this type.

- A shared knowledge plane across the layers: A common knowledge (database) plane based on all the layers information which can be accessed by all the layers when they need it. This type of interaction is very common in cross layering architectures (MobileMan [3], WIDENS [11], etc.).
- Completely new abstractions: This type almost removes the layering and considers heaps structure instead of stacks and role-based architecture, [2] is of this type.

Even cross-layer architectures can be classified how they gather different cross layer information [15]: there are architectures based on local information (from the node and its different layers) and architectures based on global information (from the node, its different layers and from neighbours). In the following we will briefly mention some of the widely discussed cross-layer architectures which may be related to autonomic networking (For more detail see [13]).

WIDENS. WIDENS (WIreless DEployable Network System) [11] has been proposed with an aim to acquire the following three main objectives at the same time (Figure 2):

Retaining inter-layer independence and peer-to-peer principles, WIDENS provides interoperability between different standards at each layer. This preserves the modularity of legacy strict layering.

A key feature is cross-layering to all protocol stacks through state (local) information and parameter mapping between adjacent layers. This mapping is beyond the legacy layering in the sense that if the local adaptation is insufficient to respond efficiently to the local performance degradation, state information and parameters mapping information to adjacent layers (from where it can then be mapped to further layers cascading through the stack if required) could improve the performance. Even the interactions between non-adjacent layers are controlled via the adjacent layers, allowing cross-layer optimisation without affecting the regular functionality of the layer. This feature rapidly reconfigures the network functions to the system constraints (bandwidth, RF) and network and application characteristics (traffic and mobility pattern) at the time of deployment.

This cross layering architectures seems a promising one where protocol optimisation is based on the local state information but it is still in the validation stage and so lacks any real measurement of efficiency especially in terms of performance.

MobileMan. The primary aim of MobileMan [3] is to exploit a MANET protocol stack's full cross-layer design. The architecture (Figure 3) along with the strict layering presents a core component, Network Status, which functions as a repository for information that network protocols collect throughout the stack. The Network Status component uniformly manages the cross-layer interaction, and respects the principle of dividing functionalities and responsibilities in layers. MobileMan achieves layer separation by standardising access to the Network Status. This reference architecture exploits the advantages of a full cross-layer design while still satisfying the layer-separation principle. This avoids duplicating efforts

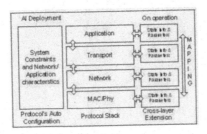

FIGURE 2. WIDENS architecture (from [11])

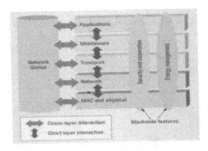

FIGURE 3. MobileMan Architecture (from [3])

to collect internal state information and leads to a more efficient system design. The approach aims to optimise overall network performance with respect to local state information by increasing local interaction among protocols, decreasing remote communications and consequently saving network bandwidth. Performance improvement verifications are yet to be published.

ECLAIR. ECLAIR [12] is an efficient cross-layer architecture for wireless protocol stacks. It is a comprehensive architecture for cross layer feedback. For the optimisation purposes it utilises local information. As shown in Figure 4, along with legacy protocol stack it consists of two main components: Optimising Sub-System (OSS), the cross layer engine. It contains many Protocol Optimisers (POs), which are the "intelligent" components of ECLAIR. The POs take input from various layers and other device entities (for example battery) and decide the optimising action to be taken.

Tuning Layers (TL) provide the necessary APIs to the POs for interacting with various layers and manipulating the protocol data structures.

Since the cross layer system is separate from the core protocol implementation, it can be easily and dynamically switched on or off, as may individual POs. There is no processing overhead on the existing stack since the optimising sub-system executes in parallel to the protocol stack. ECLAIR provides a structured approach to cross layer feedback and enables rapid deployment of new cross layer feedback algorithms.

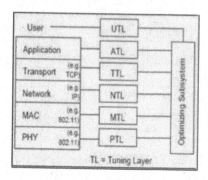

FIGURE 4. ECLAIR architecture (from [12])

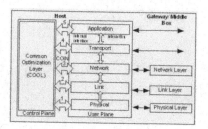

FIGURE 5. POEM Model (from [8])

Xian. Xian [1] is a generic interface for experimenting cross-layer designs with legacy 802.11 protocols. Xian can be used as a service by other network layers or system components to access information about configuration and performance of MAC/PHY layers.

POEM. POEM (Performance-Oriented Model) [8] is perhaps the first initiative towards developing a cross-layer based self-optimising protocol stack specifically for autonomic communication. For the optimisation purposes it utilises local state information. The basic design criterion is self-optimisation in a control plane issue, where the normal functions of the protocol stack should not be compromised, with added cross-layer benefits being layered on top. POEM is composed of two conceptual planes (Figure 5): the user plane for normal data flows just like without cross-layer optimisation, and the control plane for optimisation interaction flows between two protocol layers, between a protocol layer and optimisation role specific data, as well as between roles. The interactions occur through the defined Common Optimisation Interface (COIN). The Common Optimisation Layer (COOL) is responsible for offering self-optimisation services, as implemented by its Common Optimisation Protocol (COP). The system is being investigated both formally and through simulation.

The above-mentioned cross-layer architectures rely on local information and views, without considering the global networking context or views which may be

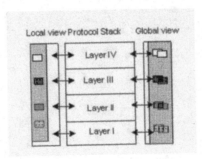

FIGURE 6. CrossTalk Architecture (from [20])

very useful for wireless networks in optimising load balancing, routing, management and even some self-behaving properties like self-organisation. Collecting and
maintaining network-wide, global statistics can be expensive, while global actions
are hard to co-ordinate. However, the effects of such systems can often be dra-
matic, and they can address problems that are difficult to detect, diagnose or
solve using purely local information. There are very few cross-layer architectures
which support both local as well as the network-wide non local view and as far
to our knowledge there are two. One is CrossTalk [20] and another is proposed in
[15]. In the following along side a brief description of CrossTalk we will go into
little deep of the architecture proposed in [15].

CrossTalk. The novel feature of CrossTalk [20] (Figure 6) is its ability to reliably establish a network-wide, global view of the network under multiple metrics.
Having such a global view, a node can use that information for local decision
processes, in conjunction with a local view containing node-specific information
contributed by each layer of the stack or system component. To keep overheads
low, no additional messages are sent: instead the local information taken from the
local view is piggybacked onto outgoing packets. Only the source of a packet is
adding its local information. Forwarding nodes do not include their information
on top. Every node inspects received packets for that information, extracts it and
adds it to its global view. This way, the global view collates numerous samples of
local information from various nodes within the network. As it is using piggyback
for global view, it is quite unlikely that any node will obtain fully accurate global
view, regardless of the frequency of data exchange. Even with an uncertain and
poor global view, however, CrossTalk has shown performance improvement in a
load balancing algorithm specifically reducing per-hop packet delay. It seems reasonable to expect such performance to be improved by improved global modelling
of the network.

Another architecture based on a combination of node-wide (local) and network-
wide (global) views has been proposed in [15]. The key distinction between this
architecture and most other cross-layer architectures is that it can not only utilise

M.A. Razzaque, S. Dobson and P. Nixon

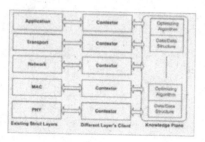

FIGURE 7. AutoComm architecture (from [15])

a node-wide local view for optimisation, but it can also use a network-wide view obtained through gossipping.

In conjunction with the existing layers, a knowledge plane is the key element of the architecture. Direct communication between layers and a shared knowledge plane across the layers are the two widely used cross-layer interactions policies [6]. Because of the improved separation and management possibilities this prefers to utilise the knowledge plane for the architecture. The following are the main elements of the architecture (as shown in Figure 7).

Alongside the normal layering support it provides the different layers' information to the knowledge plane, allowing it to maintain a local view of the node. This allows full compatibility with standards and maintains modularity, as it does not modify each layer's core functionality.

Each layer in the existing protocol stack will have a corresponding contextor, which will act as their corresponding interface between the layer and the knowledge plane. Each of these contextors will act as a "tuner" between a layer and the knowledge plane. Possible functionality for manipulating protocol data structures is built in to the contextors; no modification is required to the existing protocol stack. This facilitates incorporation of new cross layer feedback algorithms with minimum intrusion. A contextor will be responsible for reading and finally updating the protocol data structures when it is necessary.

A common Knowledge Plane (KPlane) database is maintained to encapsulate all the layers' independent information as well as the network-wide global view, which can be accessed by all layers as needed. For modularity it maintains two entities responsible for maintaining the local and global views. Interaction between different layers and the KPlane through contextors can be reactive (responding to changing context) or proactive (anticipating changes and provisioning accordingly). Generally the interactions between different layers and the KPlane are event-oriented, which suggests a reactive scheme; on the other hand, the KPlane can maintain a model of the network and act autonomously to issue its own events. This leads to improved performance if the model leads to a correct proactive adaptation, but can be detrimental if the projection is wrong. In our architecture we are considering the database with reactive interaction policies as shown in Figure

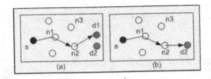

FIGURE 8. Two self-healing scenarios

7. The KPlane consists of the database and necessary optimising algorithms. The database is separated into local view and global view for isolation and management purposes, although it appears unified to clients.

Gossipping is one of the most promising data-dissemination mechanisms in peer-to-peer or distributed systems. There are number of algorithms that can be classified as reactive, proactive and periodic. In our case we propose a periodic gossiping approach, possibly with out-of-band "immediate" signalling for important changes. The gossipping service is built on top of existing TCP/UDP, and is responsible for gathering information from other nodes to generate the network-wide view at the host node. At each exchange the gossipping service chooses another node in the system (either randomly or with some weighted preference) and exchanges its local state with that node. In this architecture we will consider a gossipping exchange as an application-level event which will trigger the KPlane to take the necessary actions.

3.3.3. Sample use of cross-layer optimisation in Autonomic Networking. Possibility to cross-layer optimisation and architectures in autonomic networking may be justified by the following examples. Interestingly none of the self behaviours in autonomic computing and communications are extremely orthogonal, which means there is some dependency between them – self-healing is partly supporting self-organisation, and vice versa. Following examples based on [15] shows the possible use of cross-layering in networking in attaining self-healing or self-organising.

Consider a wireless *ad hoc* network consisting of 7 nodes (Figure 8). In the first scenario node s has a request for a service to node d1 and it is using the route s-n1-n2-d1. Using network-wide view based cross-layer architecture, all the nodes have some knowledge about their direct neighbours, so node s has knowledge about n1; n1 has about n2 and n3 and so on. If after transmission begins d1 fails, existing routing protocols would have n2 receiving the packet, determining d1 to be dead and finally sending a "node unreachable" error message to s which wastes all the resources committed to the exchange. Using a cross-layer approach, if d1 and d2 are giving almost same type of services a suitable global view would allow n2 to determine that in case of d1's failure d2 can meet the request of s. This requires making information about the service-level capabilities of a node available to the routing layer, which is facilitated by cross-layering and can easily be expressed as an optimisation algorithm. This leads to a second scenario (Figure 8 (b)) where nodes have re-organised because of the death of d1, and once n2 gets the request

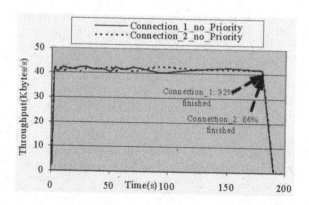

FIGURE 9. Without context-driven connections management

from s it reroutes to d2 instead of d1 and fulfils the request. With this action, cross-layer approach can conserve energy and minimises latency by eliminating the overhead required to invalidate the current route, establish a new route, and retransmit the request. Moreover, it can preserve the original route when failed node becomes available.

Context-awareness is a key concern in autonomic networks. And the possibility of context-awareness in communication through cross-layering is presented in [14]. "Context-driven traffic management by profile" is one of the implementations. Although "User" is not a layer in existing strict layering system, every layer is working directly or indirectly to provide services to user. Other than the Application Layer the rest of the layers do not understand or get the users' requirements directly and react accordingly to satisfy their demands. User requirements and corresponding context should be taken into account to enhance the user perceived QoS. The motivation for this is that the user decision could be contrary to the system decision but it could lead to improved user satisfaction. Users' preferences or priorities can be defined through a profile, based on different situations. But with the existing strict layering it is not possible to deliver this priority information to relative layers; cross-layering could be helpful and it has been shown in [14].

Consider a user in a location at the edge of WLAN's coverage with his laptop and he is downloading two files one is important FTP file and the other one is a less important music file. At the middle of downloading there is a warning about the battery and unfortunately there is no power connection nearby to charge it. In this critical situation if he continues downloading both files he might be end up with nothing finished. One can solve the problem by closing the less important connection. But with the help of cross-layer architecture like ours and integration of context-awareness it could be done automatically.

Figure 9 shows the problem when there is no priority setup, as the single file download is an atomic action so less than 100% finished means finished nothing. It shows that after 180s all the packets are dropped due to no power and resulting

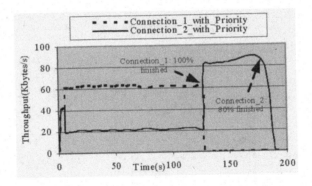

FIGURE 10. Context-driven connections management

zero download even with 92% and 86% finished for two connections respectively. Figures 10 and 11 show the results for context-driven traffic or connections management by profile. Connection-1 is the higher priority FTP download whereas Connection-2 is the lower priority music file download. Approximately at 5s simulation time there is a power warning. There is little difference between the two solutions. For the solution of Figure 10: after the warning KPlane sets the new receiver window sizes based on user level's priority. Here connection-1 has the higher priority level than connection-2 and connection-1 successfully finished downloading before (128s) power failure and after that connection-2 gets full bandwidth to download but it failed to finish before the power failure around 180s. For the solution of Figure 11: after the warning KPlane assigns the full bandwidth to the connection-1 and closes the connection-2 and connection-1 successfully finished downloading around 96s which is well before the power failure. As second solution it is closing the connection-2 just after the warning and no scope for unsuccessful finish of it, which is saving little power compared to the first solution. If we look at the graphs before and after the priority settings, the sum of the two connection's throughputs is almost same, which signifies that proportionality of the connection bandwidth share is working properly based on profile-based priority settings. Similar to previous section results these also show that cross layering could be used to utilise for context-awareness and automatically manage Internet configurations and traffic.

4. Conclusions and Future Developments

The chaotically-increasing density network of components of communications systems and the resulting growing complexity of control requires more and more distributed and self-organising structures, relying on simple and dependable elements able to collaborate to produce sophisticated behaviours. The main feature of future communication paradigms will be the ability to adapt to an evolving situation, where new resources can become available, administrative domains can

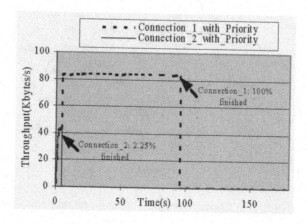

FIGURE 11. Context-driven connections management

change and economic models can vary accordingly. The move towards dynamic self-management of systems and networks – autonomic systems – without targeted human oversight provides a common framework for addressing these challenges. It is clear that the mathematical, economic and technical basis of networking must be changed radically to address the implied challenges.

Our intention in this article has been to review the state of the art in novel protocol architectures, both in terms of their technical characteristics but also – equally importantly – in terms of the business needs which these approaches serve. The notion of using a network as a profit centre is an attractive one given the decreasing marginal revenues available from simple communications alone. Autonomic communications provides an opportunity – if the current research activities are successful – to move towards this goal.

What are the next steps? Research is still needed into the ways in which context sensed at different layers, with different localities and different degrees of reliability, can be integrated into a coherent view. This "contextual fusion" is aided – but not solved – by the use of ontologies and other rich information formats. A related challenge is to provision of uncertain reasoning, so that conclusions can be drawn in a principled way even in the face of (inevitably) uncertain or contradictory information. Moreover, each adaptation that a network might make must be made in such a way as to support, rather than damage, the stability of the network as a whole. This large-scale stability in the face of small-scale changes is an essential commercial requirement for deploying autonomic networks on a large-scale basis.

Finally, the research prototypes being developed must be tested and trialed on real networks. Despite the risks, this is the only way in which we can achieve any degree of confidence in the applicability of the techniques being developed to real networks. This will in turn reveal additional requirements and opportunities to which we can apply the techniques of cross-layer adaptation and autonomic communications.

References

[1] H. Aiache, V. Conan, J. Leguay, M. Levy, *XIAN: Cross-Layer Interface for Wireless Ad Hoc Networks*. Mediterranean Ad Hoc Networking Workshop (Med-Hoc-Net), 2006.

[2] R. Braden, T. Faber, M. Handley, *From Protocol Stack to Protocol Heap — Role-Based Architecture*. First Workshop on Hot Topics in Networks (HotNets-I), 2002.

[3] M. Conti, G. Maselli, G. Turi, S. Giordano, *Cross layering in mobile Ad Hoc Network Design*. IEEE Computer, **37:2**, 2004.

[4] A. K. Dey, G. D. Abowd, *Towards a Better Understanding of Context and Context-Awareness*. Proceedings of CHI, 2000.

[5] S. Dobson, *Putting meaning into the network: some semantic issues for the design of autonomic communications systems*. Proceedings of the 1st IFIP Workshop on Autonomic Communications, **3457**, 2005, 207–216.

[6] S. Dobson, S. Denazis, A. Fernández, D. Gaïti, E. Gelenbe, F. Massacci, P. Nixon, F. Saffre, N. Schmidt, F. Zambonelli, *A Survey of Autonomic Communications*. ACM Transactions on Autonomous and Adaptive Systems, **2:1**, 2006.

[7] L. Gavrilovska, *Cross-Layering Approaches in Wireless Ad Hoc Networks*. Wireless Personal Communications, **37:3-4**, 2006, 271–290.

[8] X. Gu, X. Fu, H. Tshofenig, L. Wolf, *Towards Self-optimizing Protocol Stack for Autonomic Communication: Initial Experience*. Proceedings of the 2nd IFIP International Workshop on Autonomic Communication, **3854**, 2005, 183–201.

[9] Z. H. Haas, *Design methodologies for adaptive and multimedia networks*. IEEE Communications, **39:11**, 2001.

[10] J. Kephart, D. Chess, *The Vision of Autonomic Computing*. Computer, **36:1**, 2003, 41–50.

[11] D. Kliazovich, F. Granelli, *A Cross-layer Scheme for TCP Performance Improvement in Wireless LANs*. Globecom, 2004, 841–844.

[12] V. T. Raisinghani, S. Iyer, *ECLAIR: An Efficient Cross Layer Architecture for Wireless Protocol Stacks*. Proceedings of WWC, 2004.

[13] M. A. Razzaque, S. Dobson, P. Nixon, *Cross-layer architectures for autonomic communications*. Journal of Network and Systems Management, **15:1**, 2007, 13–27.

[14] M. A. Razzaque, P. Nixon, S. Dobson, *Demonstrating the feasibility of an autonomic communications-targeted cross-layer architecture*. Proceedings of the 14th International Conference on Advanced Computing and Communications, 2006.

[15] M.A. Razzaque, S. Dobson, P. Nixon, *A cross-layer architecture for autonomic communications*. Autonomic Networking, **4195** of LNCS, 2006, 25–35.

[16] V. Srivastava, M. Motani, *Cross-layer design: a survey and the road ahead*. IEEE Communications Magazine, **43:12**, 2005, 112–119.

[17] A. S. Tanenbaum, *Computer Networks*. 2002.

[18] D. G. Sachs *et al.*, *GRACE: A Hierarchical Adaptation Framework for Saving Energy.* Proceedings of ACEED, 2006.

[19] Q. Wang, M. A. Abu-Rgheff, *Cross-Layer Signalling for Next-Generation Wireless Systems.* Proceedings of the IEEE Wireless Communications and Networking Conference, 2003, 1084–1089.

[20] R. Winter, S. Schiller, N. Nikaein, C. Bonnet, *Cross Talk: cross-layer decision support based on global knowledge.* IEEE Communications, **44:1**, 2006.

Mohammad Abdur Razzaque, Simon Dobson and Paddy Nixon
Systems Research Group
School of Computer Science and Informatics
UCD Dublin
Belfield, Dublin 4, Ireland
e-mail: {abdur.razzaque,simon.dobson,paddy.nixon}@ucd.ie

Whitestein Series in Software Agent Technologies, 149–168

An Autonomic MPLS DiffServ-TE Domain

Rana Rahim-Amoud, Leila Merghem-Boulahia, Dominique Gaiti and Guy Pujolle

Abstract. MPLS DiffServ-TE combines the advantages of DiffServ and MPLS-TE by allowing a differentiation of services and a traffic engineering based on a fast packet switching technology. However, such MPLS DiffServ-TE network needs an efficient method for Quality of Service (QoS) guaranteeing. In addition, the management of such a network is not a simple function and could not be done manually. In fact, it would be much more economic and effective to let equipments handle a part of the tasks attributed currently to the human administrator. In this paper, a novel architecture based on a Multi-Agent System is proposed to automatically manage MPLS DiffServ-TE domains. To meet the QoS requirements, an LSP creation strategy, based on a traffic-driven approach and depending on the traffic load, is proposed. It determines when to set up a new LSP and when to forward a new traffic in an already existing one. This strategy reduces the number of LSPs and the number of signalling operations in the network. Simulation results are provided to illustrate the efficiency of our proposition.

Keywords. MPLS DiffServ-TE, Dynamic setup and re-dimension of LSPs, Multi-Agent Systems, Autonomic management.

1. Introduction

Multi-Protocol Label Switching (MPLS) [23] is a switching technology that uses short-fixed length labels to forward packets instead of using IP addresses. The classification of incoming packets is done at the entry of the domain by the Label Edge Routers (LERs) by assigning a label to the packet. Within the domain, there is no reclassification and packets are just switched by the core routers called Label Switch Routers (LSRs) according to the assigned label and the value of the incoming interface. The path between two LERs is a unidirectional path called a Label Switched Path (LSP). In recent years, there has been active research in the field of MPLS and MPLS is now a very well established networking environment. One of the most significant characteristics of MPLS is the Traffic Engineering

(TE) [20]. MPLS-TE forwards traffic flows across a network on the basis of the required resources [20]. However, MPLS does not define a new Quality of Service (QoS) architecture and cannot provide service differentiation by itself. DiffServ (Differentiated Services) [7] defines an architecture to implement scalable service differentiation in the Internet by defining multiple classes of service. In addition, as in the MPLS domain, traffic classification in the DiffServ domain is also done by edge routers by setting the Differentiated Service Code Point (DSCP) field. In the core network, there is also no reclassification; routers use the DSCP value in the IP header to select a Per-Hop Behaviour (PHB) for the packet and provide the appropriate QoS treatment. The functioning of both MPLS and DiffServ consists of 3 main steps:

1. Traffic classification,
2. Setup the path taking into account the class of the flow,
3. Traffic Forwarding following the labels.

The combination of DiffServ and MPLS presents a very attractive strategy to backbone network service providers because it provides scalable QoS and traffic engineering capabilities using fast packet switching technologies [15].

Currently, there are two solutions that have been standardised by the IETF [15]. The first one is applied to networks that support less than eight PHBs and it uses the 3 Exp (experimental) bits of the MPLS label to determine the PHB. In this case, LSPs are called E-LSPs.

The second solution is applied to networks that support more than eight PHBs. In this solution, the PHB is determined from both the label and the Exp bits and LSPs are called L-LSPs. Each solution has its advantages and its disadvantages and the use of one of them depends on the particular application scenario [19].

As networks grow rapidly and traffic conditions frequently change, the management of such a network presents many difficulties and could not be done manually. Therefore, automated management is required to minimise this complexity and to engineer traffic efficiently [9]. A few approaches are proposed in the literature to automatically manage MPLS-TE networks. Some of the proposed approaches are not designed for bandwidth guaranteed services. Thus, they cannot be used to manage MPLS DiffServ-TE domains. The other ones, intended to manage such domains, are based on a centralised management which represents a heavy and non fault tolerant solution. Therefore, we think there is a need to decentralise the network management. This led to the work presented in this paper.

In this paper, we propose a novel proactive and distributed solution in order to automatically manage MPLS DiffServ-TE domains by using Multi-Agent Systems (MAS). Our main goal is to dynamically set up and dimension LSPs depending on the actual load on the network and to meet the QoS requirements. To reach this purpose, an LSP creation strategy, based on a traffic-driven approach and depending on the traffic load, is proposed. It reduces the number of LSPs and the number of signalling operations in the network. In our proposition, we are going to consider the L-LSPs solution by using different LSPs for different classes of traffic. The result is that the physical network is divided into multiple virtual networks, one per class. These virtual networks may have different topologies and resources [26]. In this case, three virtual MPLS networks are defined for Expedited Forwarding (EF), Assured Forwarding (AF) and Best Effort (BE) classes (Figure 1). The capacity of each physical link is partitioned among these MPLS virtual networks by assigning a fixed percentage of the total link capacity to each partition. Thus each DiffServ level can be treated alone. This TE approach is called DiffServ-aware MPLS TE (DS-TE) [14]. In order to effectively control and manage the LSPs, one or more attributes can be assigned to each LSP. Two important attributes are the Setup Priority and the Holding Priority. They define eight priority levels allowing an LSP preemption between them. To guarantee the QoS, we assign the highest setup preemption priority (low preemption number) to EF, followed by AF and BE (the lowest one). So, in case of lack of available resources, LSP preemption is triggered to ensure that high priority LSPs can always be routed through relatively favourable paths. The preempted LSPs are then rerouted by their respective Ingress and Egress Label Switch Routers. The LSP preemption policy is defined in [11]. This policy can be adjusted in order to give different weight to various preemption criteria: priority of LSPs to be preempted, number of LSPs to be preempted, amount of bandwidth preempted, blocking probability. In our case, the priority is the only important criterion, the configuration adopted is then alpha=1, beta=gamma=theta=0 (for more information see the RFC 4829 [11]).

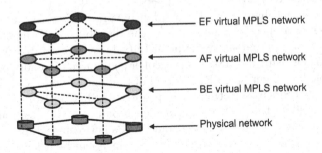

FIGURE 1. Virtual MPLS networks

The paper is organised as follows. In Section 2, we review the related work. In Section 3, we present our proposition. Next, in Section 4, we provide simulation results and evaluate the performance of the proposed strategy. The conclusion and future work are given in Section 5.

2. Related Work

To the best of our knowledge, there are only a few approaches to automatically manage MPLS-TE networks. In the following, we present some of the proposed approaches.

Traffic Engineering Automated Manager (TEAM) [25] is an automated manager for DiffServ/MPLS networks. TEAM is composed of three tools, a Traffic Engineering Tool (TET), a Measurement and Performance Evaluation Tool (MPET), and a Simulation Tool (ST). The TET is a central server which is responsible for managing the bandwidth and the routes in the network. TET is supported by the two other tools, which provide the necessary information. The TEAM architecture is very interesting. However, it is based on a central server.

The Routing and Traffic Engineering Server (RATES) [3] is another interesting tool built on centralised paradigm. RATES is a server for MPLS-TE and is developed at Bell Laboratories. RATES uses Common Open Policy Service (COPS) protocol for communicating paths and resource information to edge routers. It also uses the Open Shortest Path First (OSPF) topology database for dynamically obtaining link state information. RATES is only designed for setting-up guaranteed LSPs.

These two approaches are based on a centralised management which provides a better vision of the global network. However, it represents a heavy and non fault tolerant solution (mainly concerning the management entity). Therefore, we think there is a real need to decentralise the network control. Decentralisation of the control is obtained by allowing network components to be more independent and able to decide on actions to undertake. Furthermore, the components can, if necessary, ask for help from a human administrator or another autonomous component for the realisation of some tasks.

Another TE mechanism is MPLS Adaptive Traffic Engineering (MATE) [13]. MATE addresses a specific issue which is the distribution of the traffic flows to LSPs. MATE assumes that several explicit LSPs have been established between an ingress-egress pair using a standard protocol or configured manually. With multiple LSPs available for an egress node, the goal of the ingress node is to distribute the traffic across the LSPs by selecting the appropriate LSP. Consequently, the network utilisation as well as the network performance perceived by users are enhanced. However, for a network with N nodes and l LSPs between each pair of nodes, the total number of LSPs is of the order of lN^2, which can be a large number if the network contains many edge routers. Consequently, MATE is suitable for networks with few ingress-egress pairs. In addition, MATE is intended for traffic that does not require bandwidth reservation such as best-effort traffic.

3. Our proposition

In order to build an autonomic network, it is necessary to empower it with some essential characteristics. The characteristics that we think necessary are [22]: decentralisation, reactivity, proactivity, cooperation and adaptability. The multi-agent systems can constitute a good tool to make networks autonomic by guaranteeing these different characteristics. Indeed, an MAS consists of a set of agents that [16]:

- are able to communicate together,

- possess their own resources,
- perceive their environment (to a limited degree),
- have a partial representation of their environment and
- have a behaviour that aims to realise their purposes.

The main characteristics of agents, which make the network autonomic, are developed in the following.

3.1. Essential characteristics provided by multi-agent systems to build an autonomic network

In the following, we detail each one of the characteristics that we think necessary to build an autonomic network and we demonstrate that all these characteristics are indeed offered by the multi-agent solution.

Decentralisation. We think there is a real need to decentralise the network control. In fact, control decentralisation allows network components to decide on actions to undertake. This feature is provided by the multi-agent approach by definition. No agent possesses a global vision of the system and the decisions are taken in a totally decentralised way.

Reactivity. As the networks environment is very dynamic and is always in evolution, the router must thus be able to choose the most convenient mechanisms according to the current conditions. The multi-agent approach makes it possible because one of the basic attributes of an agent is to be situated (situadness, [8]). That is, an agent is a part of its environment. Its decisions are based on what it perceives of this environment and on its current state.

Proactivity. In an autonomic network, we should not rely only on the reactivity to control a router. In fact, a router should envisage the actions to be undertaken. This feature is also provided by the multi-agent approach. In fact, an agent can be able to set goals and to realise them by implementing plans, setting up a strategy, starting cooperation with other agents, etc.

Sociability. To guarantee end-to-end QoS between different networks, these networks should cooperate between them and reach agreements to satisfy the requirements of each of them. One of the interesting features of the multi-agent approach is its ability to distribute the intelligence between the different agents composing the system. This implies that an agent can handle some tasks individually but cannot make everything by itself. Many works concerning the concepts of negotiation and cooperation are realised and the research in this field remains very active [24]. The economic theories constituted a good source of inspiration (Contract Net Protocol, auctions, etc.) [10].

Adaptability. In order to realise its goals (accepting more traffic from a given customer, etc.), the router must be able to self-evaluate and adapt the plans to be executed. That can be allowed by using the learning feature of multi-agent systems. Researchers are interested in this feature to provide more flexibility. A part of the researches is focused on genetic algorithms [5], while the others use the reinforcement learning [12], etc.

3.2. Agent organisation and architecture

Since the MPLS functioning is based on the use of LSPs to forward packets, and the MPLS support of DiffServ is also based on the LSP, it seems that the management of LSPs is the most important need. It includes LSP dimensioning, LSP setup procedure, LSP tear-down procedure, LSP routing and LSP adaptation for incoming resource requests.

As agents have to take the convenient decisions into the MPLS domain, so the introduction of these agents will take place into the MPLS decision points. The first step of our research consists in finding the decision points of the MPLS network that are especially identified on the entry of the domain (on the LER routers) [21]. An "edge" agent will be, as a result, introduced into each LER router in the MPLS domain.

In order to effectively control and manage the network and to profit from the decentralisation feature of multi-agent systems, we decided to introduce also a "core" agent into each intermediate LSR router forming as a result an MAS. The edge and core agents have the same architecture (Figure 2), they interact and communicate together and also interact with the routers and switches in the domain. Actually, each agent is responsible for the router on which it is introduced and for the corresponding interfaces. Each agent includes two entities: the collection entity (CE) and the management entity (ME), which includes, in its turn, two sub entities: the LSP resource management entity and the LSP route management entity. In addition, the architecture contains a Data Base (DB), which is shared between the CE and the ME.

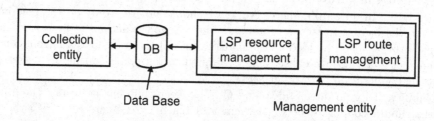

FIGURE 2. Our agent architecture

3.3. Collection entity (CE)

Each CE collects only the information concerning the interfaces of its node. The CE collects the available bandwidth of the physical links and of each LSP that goes across the node. The CE collects also the local network topology information such as the new created LSPs, if an open LSP is still in use or not, etc. The CE uses the simple network management protocol (SNMP) to collect information from the management information base (MIB) and stores them into the database.

Furthermore, the interaction between agents is done by their CEs by exchanging some of the collected information when necessary. They communicate by using specific messages that we have defined to be the simplest and the closest to network messages. Each message contains the sender agent, the receiver agent and the data of the message.

Each agent interacts with its one-hop neighbour agents and as a result gets an idea about the state of the information concerning these neighbours. Furthermore, each agent can communicate with the edge agents and ask for (or send) information. Thus, agents will be able to anticipate actions avoiding problems to occur and providing better results. This highlights the social (cooperative) and the proactive (goal-directed behaviour) features of our agents.

In this paper, we are focused on the available bandwidth because we estimate that this is the most important parameter to be treated. In fact, the available bandwidth gives a view of the current network state and allows managing appropriately the networks.

3.4. Management entity (ME)

The ME is an important part of our agent. The ME is responsible for determining when and where an LSP should be created. Indeed, the ME, which has access to the data-base, uses the stored information to take the appropriate decision. The next step performed by the ME is to automatically implement this decision. The ME contains two sub entities: the LSP route management entity and the LSP resource management entity (Figure 2).

3.4.1. LSP route management entity. The role of this entity is to set up the new LSPs on the physical network. More specifically, in case of creating a new LSP, the role of this entity is to decide, according to the load status of the network, how to select the most suitable route for the LSP to meet the QoS requirements.

Indeed, each agent maintains a threshold value, which is a criterion used to judge if a node is overloaded. This threshold value is variable and is changing dynamically according to the load status of the node. The threshold value is calculated like in [18] but with the difference that in our approach we take into account all node's interfaces and not only one. In fact, a maximum (max_{th}) and a minimum threshold (min_{th}) values are pre-determined. The threshold value of a node is initially set to max_{th} and should be always between max_{th} and min_{th}. To prevent a wrong decision on the node's load status, the threshold is calculated based on the interfaces queues occupancy and the outstanding workloads within a specific period. The outstanding workload is the mixed information of the length and the residence time of packets in an interface queue. Like in [18], when a node is considered as overloaded, its threshold value is decremented by the amount of $thresh_{dec}$. Also, if the average of queues' lengths of a node has been less than min_{th} for at least $dissolve_{th}$ seconds, which is a period long enough, the overloaded status is considered as dissolved, and the threshold value of the node returns to the initial value.

If the agent perceives that the average of the queues lengths of its node is greater than max_{th}, the agent sends a message to its one-hop neighbours informing them that its node is overloaded and cannot forward new traffics. When the over-loaded status is considered as dissolved, the agent sends a message to its one-hop neighbours informing them that its node is ready again to forward new traffics.

The threshold update algorithm is identical to the one defined in [18] with the difference that the workload is calculated as defined in the equation 3.1:

$$workload_{current} = workload_{prev} + \frac{1}{n} \sum_{i=1}^{n} que\_len_i * time_{elapsedi} \qquad (3.1)$$

where n is the number of node's interfaces. Another difference is that the test is done for the average of the queues' lengths instead of the one interface's queue length used in [18]. So, we replaced que_len in the update algorithm defined in [18] with $\frac{1}{n}\sum_{i=1}^{n} que\_len_i$.

To choose the appropriate route, the multi-agent system combines the routing information generated by a standard IP routing protocol with the local knowledge of the agent about the neighbouring load status. A signalling protocol such as RSVP-TE is then used in order to explicitly set up the LSP.

3.4.2. LSP resource management entity. The role of this entity is to find the best way to forward the incoming traffic. According to the network load and to the actual topologies, this entity decides to assign the incoming traffic to a pre-existing LSP, to re-dimension a pre-existing LSP and increase its allocated resources or to set up a new LSP. In the last case, the setting up of the LSP is done by the LSP route management entity.

To illustrate the inter-relations of these entities, we consider two possible requests: a request for LSP setup and a bandwidth request. The first request is treated by the LSP route management entity. The second one is treated by the LSP resource management entity and according to its decision, the request could be also treated by the LSP route management entity.

The agents react, when necessary, to the new environment conditions and take suitable decisions. In order to do that, we have proposed the "LSP creation strategy", which is described in the next section.

3.5. The LSP creation strategy

The main goal of this strategy is to create LSPs according to the network conditions.

In this section, we discuss the choice of the suitable approach to design the MPLS layout and the factors influencing the MAS decisions. We then present some cases of exchanging information between agents and finally, we discuss the arrival of each type of request.

3.5.1. The choice of the suitable approach to design the MPLS layout. Currently, having a physical topology, the operators are facing the challenge of designing a virtual topology to accommodate a given demand, as well as to adapt this topology to varying traffic conditions [4]. They have to find an optimal set of paths and a flow distribution over the topology.

To design the MPLS layout, there are off-line and on-line proposed approaches. Off-line approaches are based on the estimation of the traffic demand over time. According to Kodialam [17], off-line approaches are not appropriate to MPLS networks due to the high unpredictability of the Internet traffic. Thus, we avoided this solution.

On-line methods calculate paths on demand. Two different on-line approaches can be distinguished: topology-driven and traffic-driven.

In the topology-driven approach, the paths are built by a label distribution protocol according to the routing entry generated by a standard IP routing protocol [2]. A path is released only if the corresponding routing entry is deleted. In this approach, LSPs already exist before traffic is transmitted. Thus, a built path may not be used because the LSP creation was based only on routing information.

In the traffic-driven approach, the LSP is created according to the traffic information. When a new request for a flow, traffic trunk or bandwidth reservation arrives, the corresponding path is established and is maintained until the session becomes inactive [1]. In this approach, only the required LSPs are setup.

It should be noted that the available bandwidth on a physical link is equal to its maximum bandwidth minus the total bandwidth reserved by LSPs crossing it. It does not depend on the actual amount of available bandwidth on that link [26] even if the LSPs are not occupied. This means that the establishment of a non used LSP will decrease the amount of available bandwidth on the physical link and will have, as a result, bad consequences on the total MPLS network behaviour. A part of the bandwidth will be reserved without being used. Moreover, another LSP may be prevented from taking a path because of the lack of the bandwidth. In this context, the traffic-driven technology is more advantageous than the topology-driven one.

The solution, which seems the most logical and the most advantageous to design an MPLS network, is to start with an initial allocation of zero for all edges in all three virtual topologies. The arrival of the initial requests will start the process of building up the actual topologies. A topology change will take place, by consequence, when a new LSP is created or released after receiving a request. Our goal is to decide when to create a new LSP and when to allocate the new traffic on an already existing LSP. To do that we define the most important factors that may influence the MAS possible decisions.

3.5.2. The factors influencing the MAS decisions. In this section, we define the most important factors that can have an influence on the MAS decisions. These factors are: the requests and the network state.

A request can be a new bandwidth request, a request for LSP setup, a request for releasing bandwidth or a request for tearing-down an existing LSP.

The network state includes the current state of the three virtual topologies, such as the created LSPs, the existence or not of an LSP between a pair of routers, etc. The network state also includes the LSP attributes (i.e., the available bandwidth, the Setup priority, the Holding priority, etc.) and finally, the physical link attributes (i.e., the available bandwidth, the delay, etc.).

Even if the agents collect the information concerning the network state, they do not exchange all the collected information. The agents exchange only the information that they judge critical and important such as exceeding thresholds. The aim is to maintain a somewhat up-to-date vision of the network state avoiding flooding the network with useless information. In the following section, we distinguish some cases of exchanging information between agents.

3.5.3. Some cases of exchanging information between agents. In this section, we distinguish some cases where the information collected by one agent is exchanged with other agents.

The first case is when a core agent (node 2 on Figure 3a) realises that the available bandwidth on the physical link of one of its interfaces becomes lower than a well defined bandwidth level "x" (1)[1] , where "x" is a percentage of the link capacity fixed to 5%. In this case, the core agent sends this information to its one-hop neighbours and to the agents situated in the ingress routers of the LSPs which go across this link (2) (nodes 0, 1, 3, 5 and 6 on Figure 3a). If the available bandwidth of the physical link becomes higher than a well defined bandwidth level "y", where "y" is fixed to 10%, the core agent will send this information to its one-hop neighbours and to the edge agents.

Another case is when a core agent (node 5 on Figure 3b) judges that its node is overloaded (see Section 3.4.1) (1). In this case, the core agent sends a message to its one-hop neighbours (nodes 2, 4, 6 and 9 on Figure 3b) informing them that its router is overloaded and cannot forward any new traffic (2). When the router becomes less loaded, the agent sends a message to its one-hop neighbours announcing them that it is again able to forward new traffics.

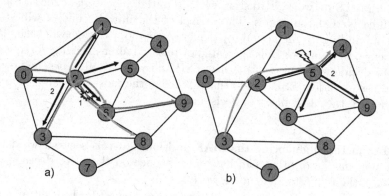

FIGURE 3. Two different exchanged information cases

[1]These values indicate the steps' succession on the Figure 3.

Other cases that we can distinguish are when an agent needs information and decides to acquire it. In fact, the proactive feature of agents can be seen with two different aspects: from the sender side and from the receiver side.

Indeed, an agent that sends information concerning a particular event does not wait that the critical threshold is being exceeded (the event is already realised) to react and inform its neighbours. In fact, the agent anticipates this critical state by taking its cautions to avoid it.

From the other side, the receiver (the neighbours) does not wait only the neighbours' messages to have an idea about the network state. The agent may decide, from time to time, to acquire the information in which it is interested and to foresee the best corresponding actions to be undertaken in the current network state (a numerical example is done in the Section 3.5.5).

3.5.4. Discussion of the arrival of each type of request. In this section, we discuss the arrival of each type of request. If a request for tearing down an already created LSP arrives at the entry of the domain, the decision is the same as if there is no MAS. The LSP will be torn down. As a result, the available bandwidth on the physical link is increased by the value of the freed bandwidth.

Consider now a request to release an allocated bandwidth. In this case, the MAS checks if the released bandwidth is equal to the total bandwidth of the LSP in which the traffic was forwarded. If so, the MAS tears down the LSP. If not, the MAS frees the reserved bandwidth and increases the available bandwidth of the corresponding LSP. Thus, the available bandwidth on the physical link remains the same.

Consider now a request to set up a new LSP. In this case, the ingress agent, which receives this request, combines the routing information generated by the routing protocol with its local knowledge about the neighbouring load status and selects the most suitable route for the LSP. The LSP is then setting up by the signalling protocol. If there is a lack of available resources and the LSP to be established has a higher priority, LSP preemption is triggered to select which lower holding preemption priority LSPs will be preempted and to preempt them. However, if it is the case of the establishment of an LSP for BE traffic (lowest setup preemption priority) and the available bandwidth is insufficient, the request is rejected.

Consider that a new bandwidth request arrives between a pair of routers demanding a certain level of QoS. The strategy for this case is defined in the diagram shown in Figure 4. In fact, the first step consists of verifying the existence of an LSP between these two routers in the corresponding virtual topology (EF, AF or BE) (2)[2]. If an LSP exists, the next step is to compare the available bandwidth of that LSP with the requested one (3). If the available bandwidth is higher than the requested one, the requested bandwidth is allocated on that LSP and its available bandwidth is reduced accordingly (4). These verifications are done by the corresponding ingress router.

---

[2]These values indicate the steps on the Figure 4.

If the available bandwidth is lower than the requested one, the multi-agent system verifies the possibility of re-dimensioning the LSP. To do that, enough bandwidth should be available in the corresponding partition of the virtual topology otherwise the MAS eliminates this possibility. If so, the requested bandwidth is compared to B_a which is the sum of the available bandwidth on the physical link and the available bandwidth on that LSP (6). This comparison is repeated for each physical link the LSP in question goes across. If the requested bandwidth is lower than or equal to B_a on one link, the comparison is done for the next one in order to reach the egress router.

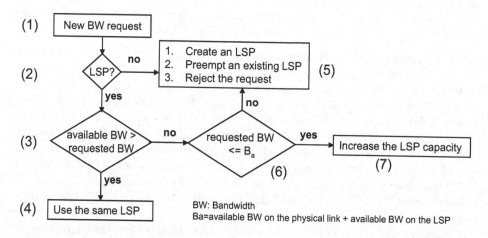

FIGURE 4. LSP creation strategy diagram for the arrival of a new bandwidth request between a pair of routers

If the requested bandwidth is lower than or equal to B_a on all physical links the LSP in question goes across, the MAS decides to increase the capacity of the LSP to be able to forward the new traffic (7). In other words, the bandwidth reserved for the LSP in question is increased by a value equal to the difference between the required bandwidth and the bandwidth available on the LSP. Consequently, the bandwidth available on the physical link is decreased by the same value. If the MAS knows that there is a loaded link or node on the path of this LSP, it avoids this path, thus reducing the allocation delay, the congestion, etc.

If the requested bandwidth is higher than B_a on one of the physical links the LSP goes across, the corresponding agent sends a message to the ingress router of that LSP informing it that it is impossible to increase the capacity of that LSP and the comparison will be interrupted for the next physical links. As a result, the MAS eliminates the possibility of re-dimensioning the LSP. If there is enough available bandwidth to satisfy the establishment of a new LSP, the MAS verifies the possibility of creating a new LSP on another physical link indicated by the routing protocol and in accordance with its own information about the loaded positions. In this case, the downstream-on-demand technique is used in order to distribute labels. If no physical link is found and it is the case of the establishment of a high priority LSP, LSP preemption is triggered to make way for the new LSP to be routed. LSP preemption is also triggered if there is a lack of available resources in the corresponding partition to establish a high priority LSP. If it is the case of the establishment of the lowest setup preemption priority LSP (BE), the request is rejected.

If there is no LSP between the pair of routers, the MAS reaction will be identical to the one where the requested bandwidth is higher than B_a.

3.5.5. Numerical example. We consider in this example the network shown in Figure 3 with a physical links' capacity equal to 400 Mbit/s. We consider that the percentages assigned to the 3 partitions EF, AF and BE are equal to 10%, 60% and 30% of the total network capacity respectively.

The LSPs in the topology are:

- $L_{0-2-6-9}$ (capacity = 70 Mbit/s, virtual topology: BE),
- L_{0-3-8} (capacity = 100 Mbit/s, virtual topology: BE),
- $L_{0-3-8-9}$ (capacity = 90 Mbit/s, virtual topology: AF),
- L_{0-3-7} (capacity = 100 Mbit/s, virtual topology: BE),
- $L_{1-2-6-8}$ (capacity = 80 Mbit/s, virtual topology: BE),
- $L_{9-6-2-3}$ (capacity = 60 Mbit/s, virtual topology: EF),
- $L_{8-6-2-3}$ (capacity = 100 Mbit/s, virtual topology: EF).

The available capacity of $L_{0-2-6-9}$ is equal to 10 Mbit/s.

The BE traffic occupies currently only 11.8% of the total network capacity. Thus, there is enough available bandwidth to forward new BE traffics.

Suppose now that a new bandwidth request arrives between R0 and R9 (1)[3]:

Source: R0, destination: R9, requested bandwidth=80 Mbit/s and the QoS demanded: BE.

The following steps are realised:

- A0 verifies the existence of an LSP between the routers R0 and R9 in the virtual topology BE (2) and finds $L_{0-2-6-9}$.
- A0 compares the available bandwidth of $L_{0-2-6-9}$ (100 Mbit/s) with the requested one (160 Mbit/s) (3). As the requested bandwidth is higher than the available bandwidth of $L_{0-2-6-9}$, A0 verifies the possibility of re-dimensioning $L_{0-2-6-9}$ by comparing the sum of the available bandwidth on the physical link and the available bandwidth on $L_{0-2-6-9}$ for each physical link it goes across (6).
- A0 verifies the link l_{0-2} (330 Mbit/s +10 Mbit/s = 340 Mbit/s > 80 Mbit/s).
- A0 sends to A2 "Is it possible to re-dimension $L_{0-2-6-9}$?"
- A2 verifies the link l_{2-6} (90 Mbit/s + 10 Mbit/s = 100 Mbit/s > 80 Mbit/s).
- A2 sends to A6 "Is it possible to re-dimension $L_{0-2-6-9}$?"
- A6 verifies the link l_{6-9} (270 Mbit/s + 10 Mbit/s = 280 Mbit/s > 80 Mbit/s).

---

[3] These values indicate the steps on the Figure 4.

- A6 sends to A2 "Re-dimensioning $L_{0-2-6-9}$ is possible".
- A2 sends to A0 "Re-dimensioning $L_{0-2-6-9}$ is possible".

As a result, the MAS decides to increase the capacity of $L_{0-2-6-9}$ to be able to forward the new traffic (7). In fact, the MAS sends a new configuration to the node (router) with the new configuration information.

The bandwidth reserved for $L_{0-2-6-9}$ is increased by the difference between the required bandwidth and the bandwidth available on $L_{0-2-6-9}$ (80 Mbit/s - 10 Mbit/s = 70 Mbit/s). Consequently, the bandwidth available on the physical links is decreased by the same value.

Available bandwidth on l_{0-2} = 330 Mbit/s - 70 Mbit/s = 260 Mbit/s.

Available bandwidth on l_{2-6} = 90 Mbit/s - 70 Mbit/s = 20 Mbit/s.

Available bandwidth on l_{6-9} = 270 Mbit/s - 70 Mbit/s = 200 Mbit/s.

A2 realises that the available bandwidth on the physical link l_{2-6} becomes lower than 5% (l_{2-6} is occupied at 95%).

A2 sends "the link l_{2-6} is overloaded" to A0, A3, A1, A5 and A6.

Elsewhere, A3 realises that node3 is overloaded and sends "node3 is overloaded" to A0, A2, A7 and A8.

We consider now that a new bandwidth request arrives between R0 and R9 in the virtual topology BE (1):

Source: R0, destination: R9, requested bandwidth = 25 Mbit/s and the QoS demanded: BE.

In this case, the following steps are realised:

- A0 verifies the existence of an LSP between the routers R0 and R9 in the virtual topology BE (2) and finds $L_{0-2-6-9}$.
- A0 knows that l_{2-6} is overloaded. As a result, A0 avoids this path and eliminates the possibility of re-dimensioning $L_{0-2-6-9}$. A0 verifies the possibility of creating a new LSP on another physical link (5). The routing protocol indicates the path (node0, node3, node8, node9).
- A0 combines this information with its own one (node3 is overloaded) and decides to avoid this path also and to create an LSP on the (node0, node2, node5, node9) which is indicated by the routing protocol to be the better path after (node0, node3, node8, node9).

By creating the LSP on this path, the MAS avoids the critical state which could exist if the LSP was created on the (node0, node2, node6, node9) or (node0, node3, node8, node9).

4. Performance evaluation

The objective of our simulations is to dynamically setup and dimension LSPs while reducing the number of LSPs and the number of signalling operations in the network. In this section, we demonstrate the performance of our proposition by comparing it to the MPLS-TE solution. We assume that in the MPLS-TE solution, the Constrained-based Shortest Path First protocol (CSPF), using the Dijkstra's shortest path algorithm [6] and the constraints for bandwidth, is used for LSP establishment. In our approach, we use this same routing protocol to create new LSPs in order to do an efficient comparison. We know how to implement the LSP preemption policy and its performance was demonstrated in [11]. Thus, in this paper, the LSP preemption will not be experimented.

The MAS decides on the route to be used regarding the routing protocol result and its own knowledge. We evaluate the performance of our proposed solution through extensive simulations on the network shown in Figure 3 using a Java simulator developed in our laboratory in Java and XML. This network includes seven edge routers and three core routers. We assume that the requests arrive one at a time at the network and only one LSP is allowed to be established per LSP request. The source and destination nodes for the requests are randomly chosen from the set of edge routers. The bandwidth request is uniformly distributed between 10 and 20 Mbits. The simulations are done for a number of requests varying from 50 to 800 requests where each request can be a new bandwidth request, a request to set up a new LSP, a request to release bandwidth or a request to tear-down an existing LSP. In addition, we have varied the physical links' capacity from 100 to 1000 Mbit/s. We repeated each simulation 7 times and we ploted the average of the obtained results. In the following we present a part of the obtained results.

Firstly, we plot the number of LSPs existing in the network after the reception of a number of requests varying from 50 to 800 requests. In Figure 5a, 5c and 5e, we present the results we obtained by setting each physical link capacity to 100, 500 and 1000 Mbit/s respectively. Simulations show that by applying our strategy we can reduce significantly the number of LSPs. Thus the number obtained by using MAS is on average reduced by a factor of 1.3 for a physical links' capacity equal to 100 Mbits/s, 1.89 and 2.11 for 500 and 1000 Mbit/s respectively. This reduction in the number of LSPs is due to the re-dimensioning decision taken by our MAS. By reducing the number of LSPs we reduce the traffic needed to control and maintain them.

The number of LSPs is not the only performance factor that our proposition enhances. Another performance factor is the number of signalling operations. In Figure 5b, 5d and 5f, we plot the number of signalling operations obtained by the MPLS-TE solution and by applying our method for a physical links' capacity equal to 100, 500 and 1000 Mbit/s respectively. The simulations show that our strategy reduces the number of signalling operations by a factor of 1.28, 2.07 and 2.5 for a physical links' capacity equal to 100, 500 and 1000 Mbit/s respectively. This reduction is attributed to the fact that, frequently, the re-dimensioned LSPs have enough space to forward the incoming traffic and there is no need to create a new LSP.

We varied the network physical links' capacity from 100 to 1000 Mbit/s. In each time, we did the simulations with the use of the MAS and without it. In Figure 6a and 6b, we plot the reducing factors for the LSPs' number and the signalling operations' number. We noted that the reducing factor is more important for the

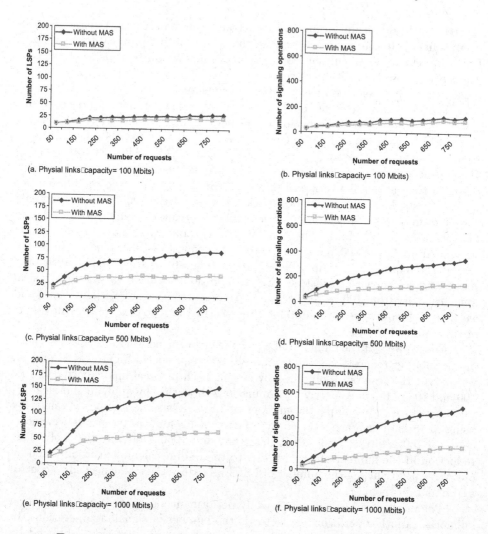

FIGURE 5. Number of LSPs and number of signalling operations in the network

higher value of the physical links' capacity. In fact, when the capacity is set to a restrictive value, only few LSPs are signalised and created in both solutions causing, as a result, a small difference between them. However, when the capacity is set to a high value, more LSPs can be signalised and created. Our strategy proposes to re-dimension the existing LSPs instead of creating new ones causing, as a result, a big difference between the results of both solutions.

In Figure 7a and 7b, we plot the reserved capacity in both solutions for a physical links' capacity equal to 500 and 1000 Mbit/s respectively. We noted that by using the MAS, the capacity reserved in the network is higher than the one reserved when the MAS is not used. This is attributed to the fact that in our strategy, when a request to release an allocated bandwidth arrives to the network and the released bandwidth is lower than the total bandwidth of the LSP in which the traffic was forwarded, the MAS frees the reserved bandwidth and increases the available bandwidth of the corresponding LSP (see Section 3.5.4). This available bandwidth is then used to forward new traffics.

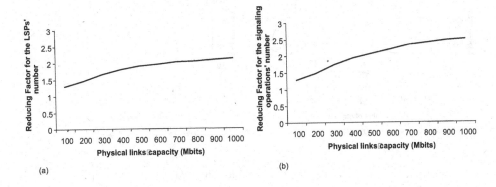

FIGURE 6. The reducing factors

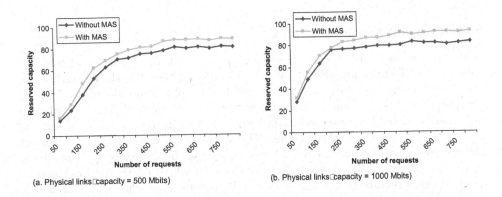

FIGURE 7. The reserved capacity

In order to verify the performance of our proposition, we calculate also the blocking rate. If there is a lack of available capacity to choose the path then the request is rejected. In Figure 8a and 8b, we plot the blocking rate for a physical links' capacity equal to 1000 and 2400 Mbit/s respectively. The number of requests is always varying from 50 to 800 requests. Simulations show that the request blocking rate obtained when using MAS is very close to the one obtained by the

MPLS-TE solution. In fact, for the low numbers of requests there is no request blocking in both solutions and for the high numbers of requests, the results are nearly identical. These results prove that our proposition does not degrade the network performance regarding the blocking rate.

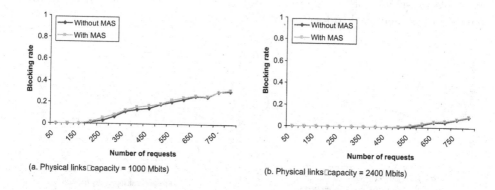

(a. Physical links capacity = 1000 Mbits)   (b. Physical links capacity = 2400 Mbits)

FIGURE 8. The blocking rate

5. Conclusion and future work

In this paper, we propose a novel architecture based on the Multi-Agent Systems to automatically manage MPLS DiffServ-TE domains. Based on the network state, the agents, being proactive, are able to make the appropriate decisions. In fact, our proposed agents do not simply react to their environment; they are able to anticipate actions. In our approach, we start with an initial allocation of zero for all edges in all three virtual topologies. The arrival of the initial requests will start the process of building up the actual topologies. Our main goal is to dynamically setup and dimension LSPs depending on the actual load on the network and to meet the QoS requirements. The challenge is to determine when to set up a new LSP and when to forward a new traffic in a pre-existing LSP. In order to do that, we propose an LSP creation strategy based on the traffic-driven approach and depending on the traffic load.

Simulation results show that our solution can significantly reduce the number of LSPs and the number of signalling operations. The reducing factors are more important for the higher values of the physical links' capacity. Furthermore, our solution does not degrade the network performance regarding the blocking rate. As future work, we are intended to define new cases of exchanging information between agents and to build experimental testbed.

References

[1] T. Anjali, C. Scoglio, J. de Oliveira, I. Akyildiz, and G. Uhl, *Optimal Policy for LSP Setup in MPLS Networks*. Computer Networks **39:2**, 2002, 165–183.

[2] G. Armitage, *MPLS: The magic behind the myths*. IEEE Communications Magazine **38:1**, 2000, 124–131.

[3] P. Aukia, M. Kodialam, P. Koppol, T. Lakshman, H. Sarin, B. Suter, *RATES: A Server for MPLS Traffic Engineering*. IEEE Network **14**, 2000, 34–41.

[4] S. Beker, D. Kofman, N. Puech, *Off Line MPLS Layout Design and Reconfiguration: Reducing Complexity Under Dynamic Traffic Conditions*. International Network Optimization Conference, INOC, 2003, 61–66.

[5] M. Berger, J.S. Rosenschein, *When to Apply the Fifth Commandment: The Effects of Parenting on Genetic and Learning Agents*. AAMAS'2004, 2004, 19–23.

[6] D. Bertsekas, R. Gallager, *Data networks (2nd ed.)* Prentice-Hall, Inc., Upper Saddle River, NJ, USA, 1992.

[7] S. Blake, D. Black, M. Carlson, E. Davies, Z. Wang, W. Weiss, *An Architecture for Differentiated Service*. RFC 2475, December 1998.

[8] R. A. Brooks, *A Robust Layered Control System for a Mobile Robot*. IEEE Journal of Robotics and Automation **2:1**, 1985, 14–23.

[9] R. Callon, *Predictions for the Core of the Network*. IEEE Internet Computing **4:1**, 2000, 60–61.

[10] E. David, R. Azulay-Schwartz, S. Kraus, *Bidders' strategy for Multi-Attribute sequential English auction with a deadline*. AAMAS-2003, 2003.

[11] J. de Oliveira, JP. Vasseur, L. Chen, C. Scoglio, *Label Switched Path (LSP) Preemption Policies for MPLS Traffic Engineering*. RFC 4829, 2007.

[12] P.S. Dutta, N. Jennings, L. Moreau, *Cooperative Information Sharing to Improve Distributed Learning in Multi-Agent Systems*. Journal of Artificial Intelligence Research **24**, 2005, 407–463.

[13] A. Elwalid, C. Jin, S. Low, I. Widjaja, *MATE: MPLS Adaptive Traffic Engineering*. IEEE INFOCOM'01, April 2001, 1300–1309.

[14] F. Le Faucheur, *Protocol Extensions for Support of Diffserv-aware MPLS Traffic Engineering*. RFC 4124, 2005.

[15] F. Le Faucheur, L. Wu, B. Davie, S. Davari, P. Vaananen, R. Krishnan, P. Cheval, J. Heinanen, *Multi-Protocol Label Switching (MPLS) Support of Differentiated Services*. RFC 3270, May 2002.

[16] J. Ferber, *Multi-Agent System: An Introduction to Distributed Artificial Intelligence*. Harlow: Addison Wesley Longman, 1999.

[17] M. S. Kodialam, T. V. Lakshman, *Minimum Interference Routing with Applications to MPLS Traffic Engineering*. IEEE INFOCOM (2), 2000, 884–893.

[18] Y. J. Lee, G. F. Riley, *A Workload-Based Adaptive Load-Balancing Technique for Mobile Ad Hoc Networks*. IEEE Wireless Communications and Networking Conference (WCNC), Georgia Institute of Technology, March 2005, 2002–2007.

[19] I. Minei, *MPLS DiffServ-aware Traffic Engineering*. Tech. report, Juniper Networks, 2004.

[20] E. Osborne, A. Simha, *Traffic Engineering with MPLS*. Cisco Press, 2003.

[21] R. Rahim-Amoud, L. Merghem-Boulahia, D. Gaiti, *Improvement of MPLS Performance by Implementation of a Multi-Agent System.* IntellComm 2005, ifip Springer, October 17-19 2005, 23–32.

[22] R. Rahim-Amoud, L. Merghem-Boulahia, D. Gaiti, *Autonomous Agents for Self-Managed MPLS DiffServ-TE Domain.* Autonomic Networking 2006, Springer - LNCS, 2006, 119–131.

[23] E. Rosen, A. Viswanathan, R. Callon, *Multiprotocol Label Switching Architecture.* RFC 3031, January 2001.

[24] T. Sandholm, D. Levine, M. Concordia, P. Martyn, R. Hughes, J. Jacobs, D. Begg, *Changing the Game in Strategic Sourcing at Procter & Gamble: Expressive Competition Enabled by Optimization.* Special issue on the 2005 Edelman award competition **36:1**, 2006, 55–68.

[25] C. Scoglio, T. Anjali, J. de Oliveira, I. Akyildiz, G. Uhl, *TEAM: A Traffic Engineering Automated Manager for DiffServ-Based MPLS Networks.* IEEE Communications Magazine **42:10**, 2004, 134–145.

[26] X. Xiao, A. Hannan, B. Bailey, L. Ni, *Traffic engineering with MPLS in the Internet.* IEEE Network Magazine, 2000, 28–33.

Acknowledgment

This work has been partially funded by the Champagne-Ardenne Regional Council and the European Social Fund.

Rana Rahim-Amoud
ICD/ LM2S, CNRS FRE 2848, UTT
12, rue Marie Curie
BP 2060 10 010 Troyes Cedex, France
e-mail: rana.amoud@utt.fr

Leila Merghem-Boulahia
ICD/ LM2S, CNRS FRE 2848, UTT
12, rue Marie Curie
BP 2060 10 010 Troyes Cedex, France
e-mail: leila.boulahia@utt.fr

Dominique Gaiti
ICD/ LM2S, CNRS FRE 2848, UTT
12, rue Marie Curie
BP 2060 10 010 Troyes Cedex, France
e-mail: dominique.gaiti@utt.fr

Guy Pujolle
LIP6, Université de Paris 6
8 rue du Capitaine Scott
BP 2060 75015 Paris, France
e-mail: guy.pujolle@lip6.fr

Whitestein Series in Software Agent Technologies, 169–190

Game Theoretic Framework for Autonomic Spectrum Management in Heterogeneous Wireless Networks

Jie Chen, Miao Pan, Kai Yu, Yang Ji and Ping Zhang

Abstract. In order to enable the autonomic network resource management, we resort to game theory to facilitate the spectrum sharing between heterogeneous wireless networks. We set the spectrum sharing objective to improve spectrum efficiency and maximise network revenue. Based on this goal, we propose a revenue sharing bargain game to model the spectrum sharing behaviour of intra-operator radio access networks (RANs) and a price bargain game to model the spectrum trading behaviour between inter-operator RANs. Furthermore, we adopt multi-agent architecture to implement the spectrum sharing schemes autonomously. Both bargaining mechanisms are analysed using bargaining game theory, and consequently the implementations are refined and simplified based on the analysis. Simulation results show that the proposed mechanisms outperform the conventional fixed spectrum management method in network revenue, spectrum efficiency and call blocking rate.

Keywords. Autonomic spectrum management, Game theory, Multi-Agent.

1. Introduction

As one of the important resources in wireless networks, the spectrum is regulated by governmental agencies and is allocated to license holders or operators on a long term basis for large geographical regions. However, the fixed allocation of spectrum might not always provide the optimal spectrum efficiency in the future reconfigurable multi-radio environment, with traffic loads varying spatially and temporally. In addition, the growing demand for wireless mobile multimedia services, and hence increased traffic capacity, highlights a need for more autonomy in terms of spectrum allocation and coordination among multi-Radio Access Network (RANs). How to autonomously maximise the spectrum efficiency to provide as many services as possible and maximise the profits of networks poses a great challenge.

Recently, considering the scarcity and high economic value of spectrum, worldwide research efforts have been made in Dynamic Spectrum Management (DSM) [1, 2, 3, 4]. Several mechanisms have been proposed to enhance the efficiency of spectrum utilisation, e.g., the European research project DRiVE and OverDRiVE [1, 2, 3]. With the rapid development of software defined radio (SDR) and the introduction of reconfiguration concepts [4], DSM no longer remains a utopia but becomes a reality. However, among all the contributions above, as well as other works, little efforts have previously been made on the potential for win-win solutions between RAN's spectrum efficiency and the RAN's profits.

As a solution to DSM, spectrum sharing between RANs is considered in this paper. We propose that the spectrum should be traded autonomously and periodically between the RANs and the revenue, earned by the traded spectrum, should also be shared and negotiated between the RANs. The negotiation procedure can be modelled as an infinite-horizon alternating offer bargaining game with perfect information. Based on bargaining game theory [5], the resulting bargaining game has unique equilibrium, and the negotiation procedure can reach an agreement immediately. Thus, the bargaining is of great efficiency. Since the RANs are needed to be perfectly informed about each other RAN's private key information, this approach is applied more appropriately between intra-operator RANs than between inter-operator RANs.

In order to facilitate the spectrum sharing between RANs belonging to different operators, we further propose a bargaining scheme to negotiate over the price of traded spectrum block. This negotiation procedure can be modelled as an infinite-horizon bargaining game with one-sided uncertainty. Since this game has unique sequential equilibrium, the optimised implementation of the negotiation procedure is proposed.

Through extensive simulations, the results show that the proposed schemes are effective in improving the spectrum efficiency while increasing the RANs' revenue, either inter-operator or intra-operator.

The rest of this paper is organised as follows. We begin in Section 2 by describing the related works on dynamic spectrum sharing. Section 3 presents the system architecture for spectrum trading. Based on bargaining game theory, the spectrum sharing schemes between intra-operator RANs and inter-operator RANs are proposed in Section 4 and Section 5, respectively. We conduct a plenty of simulations to evaluate the proposed schemes in Section 6, and provide an agent implementation scheme in Section 7. Finally, Section 8 concludes the whole paper.

2. Related Works

Up to date, the spectrum sharing between networks has been regulated via fixed spectrum management (FSM) among different systems or centralised allocations between different base stations of a system in cellular networks. In ad-hoc networks, only the interference issues in the ISM band have been investigated focusing mostly on the coexistence of WLAN and Bluetooth networks.

Jing *et al.* [8] proposed the common spectrum coordination channel (CSCC) etiquette protocol for coexistence of IEEE 802.11b and 802.16a networks. In this scheme, each node is assumed to be equipped with a cognitive radio and a low bit-rate, narrow-band control radio. The coexistence is maintained through the coordination of these nodes with each other by broadcasting CSCC messages. Each user determines the channel it can use for data transmission such that interference is avoided. In case channel selection is not sufficient to avoid interference, power adaptation is also deployed. This scheme is efficient in improving throughput.

In addition to the competition for the spectrum, competition for the users was also considered in [9]. In this work, a central spectrum policy server (SPS) was proposed to coordinate spectrum demands of multiple operators. In this scheme, each operator bids for the spectrum indicating the cost it will pay for the duration of the usage. The SPS then allocates the spectrum by maximising its profit from these bids. The operators also determine an offer for the users who, in turn, select an operator to serve a given type of traffic. When compared to a case where each operator is assigned an equal share of the spectrum, the operator bidding scheme achieves higher throughput leading to higher revenue for the SPS, as well as a lower price for the users according to their requirements. This work opened a new perspective by incorporating competition for users as well as the spectrum in future wireless networks.

Marias [10] proposed a distributed spectrum sharing scheme for wireless Internet service providers (WISPs) that share the same spectrum, where a distributed QoS based dynamic channel reservation (D-QDCR) scheme is used. The basic concept behind D-QDCR is that a base station (BSs) of a WISP competes with its interferer BSs according to the QoS requirements of its users to allocate a portion of the spectrum. Similar to the CSCC protocol [8], the control and data channels are separated. The basic unit for channel allocation in D-QDCR is called Q-frames. When a BS allocates a Q-frame, it uses the control and data channels allocated to it for coordination and data communication between the users. The competitions between BSs are performed according to the priority of each BS depending on the data volume and QoS requirement. Moreover, various competition policies are proposed based on the type of traffic a user demands.

Many works also focus on intra-network spectrum sharing, where the secondary users of a network try to access the available spectrum without causing interference to the primary users.

A cooperative local bargaining (LB) scheme was proposed in [11] to provide both spectrum utilisation and fairness. Local bargaining is performed by constructing local groups according to a poverty line that ensures a minimum spectrum allocation to each user and hence focuses on fairness of users. The localised operation provides an efficient operation between a fully distributed and a centralised scheme. Moreover, local bargaining has close performance compared with centralised graph colouring approach at a reduced complexity.

Another approach that considers local groups for spectrum sharing was provided in [12], where a clustering algorithm was proposed such that each group selects a common channel for communication, and distributed sensing and spectrum sharing is provided through this channel. Moreover, if this channel is occupied by a primary user at a specific time, the nodes reorganise themselves to use another control channel. The performance evaluations show that the distributed grouping approach outperforms common control channel approaches especially when the traffic load is high.

Brik et al. [13] presented a centralised solution for intra-network spectrum sharing with fixed infrastructure, which is called dynamic spectrum access protocol (DSAP). The DSAP enables a central entity to lease spectrum to users in a limited geographical region. DSAP consists of clients, DSAP server, and relays that relay information between server and clients that are not in the direct range of the server. Moreover, clients inform the server their channel conditions so that a global view of the network can be constructed at the server. By exploiting cooperative and distributed sensing, DSAP servers construct a Radio Map. This map is used for channel assignments which are leased to clients for a limited amount of time.

In addition, game theory has also been exploited for performance evaluation of spectrum access schemes. For instance, Nie et al. [14] provided the comparison between cooperative and non-cooperative approaches through game theoretical analysis.

So far, the existing various spectrum sharing schemes did not consider the spectrum sharing method between RANs belonging to the same operator or different operators, which must take into account some other aspects, such as economic influence and information acquirement. Consequently, this work focuses on the intra- and inter-operator spectrum sharing schemes, and takes revenue, spectrum efficiency and call blocking rate into consideration.

3. System Architecture

3.1. Spectrum Sharing Architecture

The architecture proposed for spectrum sharing is shown in Figure 1. Virtual Spectrum Market (VSM), which originates from the definition of spectrum pool [4], is a logical spot where RANs could trade spectrum with each other. In the VSM, if some RANs have surplus spectrum resource to the service requirements, they can lease the extra spectrum out to maximise the spectrum efficiency and the profits. On the other hand, if some RANs lack of spectrum to satisfy the temporarily increasing services, they become the consumers of the VSM. In order to satisfy as many service demands as possible to decrease call blocking rate and make more profits, these RANs will try to rent spectrum from others to proceed with their service provisioning.

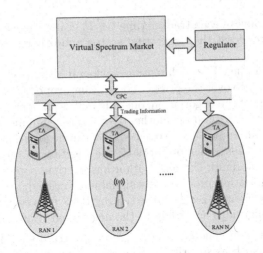

Figure 1: The spectrum sharing architecture in reconfigurable system.

Once there is a market, there should be policies to regulate the operation of trading behaviours [15]. Regulator in Figure 1 is a repository of these principles, e.g., trading regulations, hostile competition bans and so on.

In the VSM, all trading RANs should know some necessary trading information, for instance, which RANs will lease or rent spectrum, what frequencies are spare at the moment, and so on. One promising solution is to utilise the concept of common pilot channel (CPC) [16] to inform RANs the trading information. With the help of CPC, the involved RANs can acquire the preliminary information before trading and exchange transaction opinions in trading.

Furthermore, to facilitate the spectrum trading, we propose to divide the entire spectrum block into spectrum trading units (STU) in a size of the smallest channel in a fixed service channel raster for a given band. The trouble in controlling the interference is ignored in the proposed spectrum sharing schemes.

As mentioned in other works [2, 3], the spectrum sharing is a periodical and proactive operation and the period is in hour scale. In the following, denote T by the set of spectrum sharing time. To fulfil the spectrum trading, each RAN is supposed to be attached with an intelligent Trading Agent (TA). TAs take charge of the trading and make a series of important decisions, e.g., when to lease/rent the spectrum, how much spectrum to lease/rent, at what price to lease/rent the spectrum, how to deal with the profits brought by spectrum trading, etc.

3.2. General Spectrum Sharing Process

In general, the spectrum sharing process can be depicted as follows:

- At the beginning of each trading time $t \in T$, either the leasing RANs or renting RANs broadcast their spectrum status and other necessary information in CPC.
- By monitoring the CPC, each renting/leasing RAN selects one leasing/renting RAN to negotiate over the price or some other properties of the trading STU. At every stage of the negotiation, one side makes an offer, while the other side decides to accept or propose another offer if reject. Should one side accept the other's offer, an agreement is reached and the negotiation is over. Whereas, a rejection by this trader prompts the negotiation to the next round. Note that on the one hand, the negotiation has no time constraint, on the other hand the negotiation process costs signaling overhead.
- When an agreement is reached, the renting RAN uses the STU at period $t+1$ and fulfils obligations according to the negotiated results.

Note that this is a general description on spectrum sharing procedure. Due to various reasons such as security, self-interest, the information shared between inter-operator RANs are less than that between intra-operator RANs. Thus the spectrum sharing procedure between intra-operator RANs is different from that between inter-operator RANs. The following two sections propose the detailed solutions for the two circumstances.

4. Spectrum Sharing Bargaining Scheme with Perfect Information

When all RANs belong to the same operator, they can know the necessary information for spectrum trading, so that the negotiations between renting RAN and leasing RAN can be considered as a bargaining game with perfect information [5]. We call this scheme as intra-operator spectrum sharing bargaining scheme. As one of the most classic models in that game, infinite-horizon alternating offer bargaining game with perfect information is adopted for the revenue-sharing in this scheme and related implementation procedures are investigated in this part as well.

4.1. Profits of RANs

The spectrum sharing scheme runs periodically. At time t, a pair of TAs, constituting the traders of the spectrum trading, negotiates over the division of the revenue earned by a STU providing service for renting RAN's subscribers at time $t+1$. Say the agreement is pair (x_r, x_l), $x_r + x_l = 1$. Then, when the STU earns revenue at time $t+1$ in practice, the revenue will be shared between the leasing TA and the renting TA based on the negotiated ratio.

Due to the proactive features of the scheme, practical traffic demands present some deviations from predicted traffic demands to some extent in general. Provided that practical load is less than predicted load, renting TA's investment on trading spectrum at period t may bring no profits at period $t+1$. To guarantee the profits of leasing TAs, a simple risk capital mechanism is embodied in the scheme. That is, when renting TA trades with leasing TA at time t, it should deliver additionally a small advance payment to leasing TA beforehand. The advance payment, expressed as γ, is non-refundable.

Denote λ by the profit earned by a renting RAN utilising a STU to provide services for one period. Then the net profits earned by leasing RAN and renting RAN according to the spectrum trading can be expressed as:

$$\begin{cases} profit_l & = & x_l \cdot \lambda + \gamma \\ profit_r & = & x_r \cdot \lambda - \gamma \end{cases} \tag{4.1}$$

4.2. Spectrum Sharing Bargaining with Perfect Information

4.2.1. General Description.
Infinite-horizon alternating offer bargaining game with perfect information was used as a solution to the problem that two bargainers are negotiating over the division of a "cake" of size 1 [5]. We introduce this model into the intra-operator spectrum sharing scheme and analyse the negotiation result. At first, we give the definition of the spectrum bargaining game.

Definition 4.1. The intra-operator spectrum sharing bargaining game with perfect information is the following extensive game:

Players: Leasing TA (TA_l for short) and renting TA (TA_r for short), say l and r constituting set $I = l, r$.

Terminal history: Every sequence of the form $(\mathbf{x}_1, N, \mathbf{x}_2, N, \dots, \mathbf{x}_n, Y)$ for $n \geq 1$, and every infinite sequence of the form $(\mathbf{x}_1, N, \mathbf{x}_2, N, \dots)$, where each $\mathbf{x}_n$ is a proposed division vector of the profit earned by STU. All the division vectors consist of the following set:

$$\mathbf{X} = \left\{ (x_l, x_r) \in \mathbf{R}^2 : x_l + x_r = 1, x_i \geq 0, i \in I \right\}$$

Player function: $P(\varnothing) = l$ (TA_l makes the first offer), and

$$P(\mathbf{x}_1, N, \dots, \mathbf{x}_n) = P(\mathbf{x}_1, N, \dots, \mathbf{x}_n, N) = \begin{cases} l, & n \text{ is even} \\ r, & n \text{ is odd} \end{cases}$$

Preferences: TA_l's and TA_r's payoff to the terminal history $(\mathbf{x}_1, N, \mathbf{x}_2, N, \dots, \mathbf{x}_n, Y)$ is

$$\begin{aligned} u_l & = & \delta_l^{n-1} \cdot (x_{l,n} \cdot \lambda + \gamma), & \delta_l \in (0,1) \\ u_r & = & \delta_r^{n-1} \cdot (x_{r,n} \cdot \lambda - \gamma), & \delta_r \in (0,1), \end{aligned}$$

and their payoff to every infinite terminal history is zero, where $\delta_i \in (0,1)$ is called the discount factor or the bargaining patience.

In detail, the bargaining procedure is as follows. The players can take actions only at rounds in the infinite set $N = \{1, 2, \dots\}$. In the first round, the TA_l announces an agreement $\mathbf{x}, (\mathbf{x} \in \mathbf{X})$. If the TA_r accepts the offer, then bargaining game is over, and the agreement is applied. If the TA_r rejects the offer, the game proceeds to the next round. In this round, the TA_r proposes an agreement, which TA_l can choose to accept or reject. The bargaining continues in this manner. Whenever an offer is rejected, the bargaining proceeds to the next round, in which it is the rejecting TA's turn to announce a new offer. There is no limit on the number of rounds. At all rounds, both TA_r and TA_l know all previous moves of each other and their own. In practice, the bargaining procedure takes time, signalling overhead and other resource. Therefore, the players' payoffs are discounted after each round negotiation according to the discount factor δ_r for the TA_r and δ_l for the TA_l, with $0 < \delta_r, \delta_l < 1$.

The discount factor can be explained as bargaining patience, which depends on the RAN's spectrum demand. To be specific, at current time t, if TA_r predicts that his loads increase at $t + 1$, which means he is in more urgent need of the spectrum, he will be less patient and bargain less. On the contrary, if he predicts his loads decrease at $t + 1$, his bargaining patience factor δ_r will increase. Correspondingly, if TA_l predicts that his loads increase at $t+1$, which means his surplus spectrum is shrinking, he will be more patient and bargain more. And if he predicts his loads decrease at $t + 1$, which means he has more available spectrum to lease, his bargaining patience factor δ_l will decrease.

This implies the following requirement on the derivative of both leasing and renting RAN's discount factors:

$$\frac{\partial \delta_i(\Psi_i)}{\partial \Psi_i} < 0, \quad (\Psi_i > 0, i \in I), \tag{4.2}$$

where Ψ_i represents the required STU number of renting RAN or the surplus STU number of leasing RAN.

Considering the practical cases, when Ψ_i is small, the discount factor is close to 1 and has to decrease for increasing Ψ_i, but the discount factor is not much sensitive to the increase of Ψ_i, which means that the decreasing speed is slow. Meanwhile, when Ψ_i is large, the discount factor is close to 0 and is not sensitive to the increase of Ψ_i as well. Therefore, the discount factor should also satisfy the following requirements:

$$\exists \xi \in (0,1), \quad \begin{cases} \Psi_i < \xi: & \frac{\partial^2 \delta_i(\Psi_i)}{\partial \Psi_i^2} < 0 \\ \Psi_i > \xi: & \frac{\partial^2 \delta_i(\Psi_i)}{\partial \Psi_i^2} > 0 \end{cases} \tag{4.3}$$

In this work, the discount factor is modeled as inverse sigmoid curves assuring the validity of conditions (4.2), (4.3). We propose the following analytic expression for it:

$$\delta_i(\Psi_i) = \frac{1}{1 + (\Psi_i/K_i)^\nu}, \quad \nu \geq 2 \tag{4.4}$$

where K_i is the number of total STUs originally assigned to RAN i. In addition, the conclusions we derived about the properties of discount factor are quite general and do not depend on this particular choice. They are valid for every function that satisfies Eqs.(4.2), (4.3).

Note that at the same period t, TA_r and TA_l may bargain for many rounds, and in each round their discount factors, δ_r and δ_l, remain invariable.

4.2.2. Equilibrium of the Intra-operator Spectrum Sharing Bargaining Game with Perfect Information. Similar to the proof in [5], the intra-operator spectrum sharing bargaining game with perfect information has the unique equilibrium, which is expressed as follows:

$$(x_l^*, x_r^*) = \left(\frac{\delta_r \cdot (1 - \delta_l)}{1 - \delta_r \cdot \delta_l} - \frac{\gamma}{\lambda}, \frac{1 - \delta_r}{1 - \delta_r \cdot \delta_l} + \frac{\gamma}{\lambda} \right) \tag{4.5}$$

It's obvious that if TA_r's patience factor δ_r decreases, TA_l will get more proportions of the profits and if TA_l's patience factor δ_l decreases, TA_r will get more proportions of the profits. The more patient they are, the more profits they will receive. In summary, when a trader learns his opponent's card beforehand, the final result concerning the allocation of the revenue-sharing completely depends on their patience extent and can be drawn immediately. Since the bargaining game with perfect information has a unique equilibrium, there is no need for TA_l and TA_r to negotiate over the revenue sharing ratio. When the discount factor of both traders are common knowledge, TA_l and TA_r can calculate directly the equilibrium sharing ratio according to Eq. (4.5). Therefore, the spectrum bargaining with perfect information is of high efficiency. Consequently, the implementation of the intra-operator spectrum sharing scheme can be simplified.

4.2.3. Implementation of Spectrum Sharing Bargaining Scheme with Perfect Information. The spectrum sharing bargaining scheme runs periodically. In each period, TA takes four main steps: traffic load predicting, discount factor calculating, discount factor exchanging and decision making. The overall procedure is illustrated in Figure 2.

Step 1: Load Predicting

Based on predefined load prediction algorithm, TA predicts the traffic load at next time $t + 1$. Then, TA calculates the spectrum requirement, S_p, according to the predicted traffic load. Furthermore, the spectrum usage ratio, r_s, can be calculated as:

$$r_s = S_p / K_i \tag{4.6}$$

where K_i is the number of total STUs originally assigned to RAN i. Thus, all the RANs can be classified into two categories with respect to the spectrum usage ratio.

- If $r_s > Th$, according to load prediction, current RAN needs to rent some spectrum from the others in the next period and his TA now acts as the renting TA.
- If $0 < r_s < Th$, according to load prediction, current RAN has some spare spectrum to share with the other in the next period and his TA now acts as the leasing TA.

Here, the variable Th is defined as trading threshold to preclude waste of spectrum and signaling overhead. In addition, Th can also be used for resource reservation.

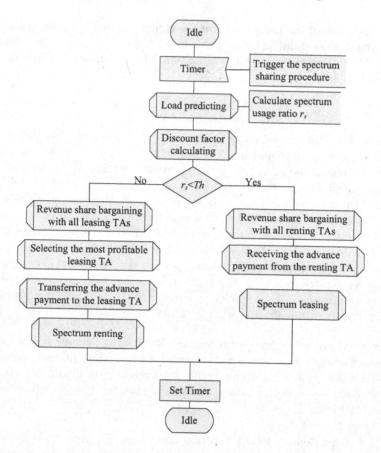

Figure 2: The procedure of Intra-operator Spectrum Sharing Bargaining Scheme.

Step 2: Discount Factor Calculating

Based on the predicted spectrum usage, TA calculates its own discount factor according to Eq. (4.4). And the discount factor remains constant in this period negotiation.

Step 3: Discount Factor Exchanging

Each TA_l exchanges its own discount factor with each TA_r.

Step 4: Decision Making

TA_l will weigh all the potential bargaining contracts and choose the most profitable TA_r to do the spectrum trading.

Note that this scheme is practical when the trading occurs between intra-operator RANs, because all RANs within the same operator can share their discount factor and other necessary information. Moreover, these RANs are willing to share the revenue. However, when the spectrum trading occurs between inter-operator RANs, the circumstance becomes complex. These RANs are more willing to lease or rent the spectrum at certain price instead of sharing the revenue. They will bargain over the price of spectrum under incomplete information.

Consequently, we further propose a spectrum sharing scheme for inter-operator RANs in the following Section.

5. Spectrum Sharing Bargaining Scheme with Incomplete Information

When the spectrum sharing occurs between inter-operator RANs, we propose that the renting RAN pays the leasing RAN for each rented STU. Thus, at what price the STU should be traded is negotiated between the two RANs. When an agreement is reached, the renting RAN pay the leasing RAN the negotiated price to use the spectrum. Different from intra-operator spectrum sharing scheme, this negotiation proceeds under incomplete information.

We model the bargaining between two inter-operator RANs as infinite-horizon bargaining game with one-sided incomplete information [7]. Furthermore, we predict the behaviour of the negotiation and derive the result of the bargaining game based on the theoretic analysis.

5.1. Spectrum Sharing Bargaining with One-Sided Uncertainty

5.1.1. General Description. Without loss of generality, we consider the spectrum bargaining game in which the leasing RAN makes all of the offers and has incomplete information about the renting RAN's valuation on each STU. The leasing RAN has its own valuation s on each STU and has discount factor δ_l, both of which are common knowledge. The renting RAN's valuation, b, is private information and known only to himself. The leasing RAN knows the renting RAN's valuation distribution density, $f(b)$, on the interval $[\underline{b}, \overline{b}]$, where $\underline{b} \geq s$, for any renting RAN with $b < s$ would not enter negotiations. The renting RAN's discount factor, δ_r, is common knowledge to both leasing and renting RANs. Firstly, we give the definition of this inter-operator spectrum sharing game.

Definition 5.1. The inter-operator spectrum sharing bargaining game with incomplete information is the following extensive game:

Players: Leasing TA (TA_l for short) and renting TA (TA_r for short), say l and r.

Terminal histories: Every sequence of the form $(p_1, N_2, p_3, N_4, \ldots, p_{n-1}, Y_n)$ for $n \geq 1$, and every infinite sequence of the form $(p_1, N_2, p_3, N_4, \ldots)$, where each p_r is a price proposed by TA_l, and N, or Y is the strategy set of TA_r.

Player function: $P(\varnothing) = l$ (TA_l makes the first offer), and

$$\begin{cases} P(p_1, N_2, p_3, N_4, \ldots, N_n) = l, & n \text{ is even} \\ P(p_1, N_2, p_3, N_4, \ldots, p_n) = r, & n \text{ is odd} \end{cases}$$

Preferences: TA_l's and TA_r's payoff to the terminal history $(p_1, N_2, p_3, N_4, \ldots, p_{n-1}, Y_n)$ is

$$
\begin{aligned}
u_l &= \delta_l^{n-1} \cdot (p_{n-1} - s), \quad \delta_l \in (0,1) \\
u_r &= \delta_r^{n-1} \cdot (b - p_{n-1}), \quad \delta_r \in (0,1),
\end{aligned}
$$

and their payoff to every infinite terminal history is zero.

At every stage $n \geq 1$, of the game, the TA_l chooses an optimal price schedule p_n, and TA_r accepts the offer if his valuation is greater than some indifference valuation $b_n(p_n)$ (if the TA_r's valuation is less than $b_n(p_n)$, he is better off holding out for lower offers in the future). Thus, a rejection by the TA_r indicates to the TA_l that the TA_r's valuation is less than b_n.

A few comments are in order. First, only one side of each pair of traders knows the other side's valuation, which in turn determines the result of the following negotiation. Second, as the negotiation process costs time and signaling overhead, both sides of the traders are not willing to take long time to bargain over the payment. Thus their payoffs in the subsequent rounds are discounted.

5.1.2. Sequential Equilibrium of the Inter-operator Spectrum Sharing Bargaining Game with Incomplete Information.

This part finds the sequential equilibrium of the inter-operator spectrum sharing bargaining game. Sequential equilibrium is a refinement of Nash Equilibrium for extensive form games due to David M. Kreps and Robert Wilson [6]. A sequential equilibrium specifies not only a strategy for each of the players but also a belief for each of the players. A belief gives, for each information set of the game belonging to the player, a probability distribution on the nodes in the information set. A profile of strategies and beliefs is called an assessment for the game. Informally speaking, an assessment is a sequential equilibrium if its strategies are sensible given its beliefs and its beliefs are sensible given its strategies.

Formally, we give a definition of sequential equilibrium of the inter-operator spectrum sharing bargaining game.

Definition 5.2. The sequential equilibrium behaviour of the TA_l and TA_r is a sequence of prices for TA_l and a sequence of indifference valuations for TA_r that satisfies the following property:

Sequential Rationality: The TA_l's future offers $p_1, p_2, \ldots$ are chosen to maximise the payoff of TA_l given by the TA_r's future indifference valuations $b_1, b_2, \ldots$, which are chosen so that TA_r is indifferent between accepting p_n now or waiting one round and accepting p_{n+1} next round:

$$
b_n - p_n = \delta_b \cdot [b_n - p_{n+1}], \quad \forall n \geq 0
$$

The players' equilibrium behaviour is determined by solving a dynamic programming problem in which the TA_l chooses the offer that maximises his present value of current and future gains, given his knowledge of the TA_r's valuation, and subject to the constraint that the TA_r will accept the offer only if his valuation is sufficiently high that he is better off accepting now than waiting for lower prices in the future. Namely, in round n in the bargaining game, the TA_l chooses p_n to maximise his expected gain $u_l^{(n)}(s, b_n)$ given that the TA_r's valuation is distributed on $[\underline{b}, b_n]$ with distribution density being $f(b)$. The dynamic programming problem can be expressed as:

$$\max_p u_l^{(n)}(s, b_{n-1}) = \max_p [\int_{b_n}^{b_{n-1}} (p - s) \cdot f(b)db$$

$$+ \delta_l \cdot \int_{\underline{b}}^{b_n} u_l^{(n+1)}(s, b_n) \cdot f(b)db] \tag{5.1}$$

subject to

$$b_n - p = \delta_r \cdot (b_n - p_{n+1})$$

As proved by Fudenberg *et al.* [7], we have the following proposition.

Proposition 5.3. *If $\underline{b} \geq s$, the pre-defined spectrum bargaining game has sequential equilibrium and the equilibrium is unique and weak-Markov.*

As a special case, we consider that the TA_r's valuation is uniformly distributed on $[\underline{b}, \bar{b}]$, i.e.,

$$f(b) = \frac{1}{\bar{b} - \underline{b}}, \quad b \in [\underline{b}, \bar{b}].$$

Under this circumstance, we present the sequential equilibrium, which is the solution to Eq. (5.1). The following lemma describes the equilibrium.

Lemma 5.4. *When $\underline{b} \geq s$ and the TA_r's valuation is uniformly distributed on $[\underline{b}, \bar{b}]$, then the TA_l's equilibrium price p_n in round n, his expected profit u_l, and the TA_r's indifference valuation b_n in round n are given by*

$$p_n = c \cdot (\bar{b} - s) \cdot d^{n-1} + s$$
$$u_l = \tfrac{1}{2} \cdot c \cdot \frac{(\bar{b}-s)^2}{\bar{b}-\underline{b}} \tag{5.2}$$
$$b_n = (\bar{b} - s) \cdot d^{n-1} + s$$

where $d = c/(1 - \delta_r + \delta_r \cdot c)$ and $c(\delta_l, \delta_r)$ is defined implicitly by

$$c = \frac{(1 - \delta_r + \delta_r \cdot c)^2}{2 \cdot (1 - \delta_r + \delta_r \cdot c) - \delta_l \cdot c}.$$

The equations for c and d above can be solved simultaneously to yield

$$c = \frac{1 - \delta_r}{1 - \delta_r + \sqrt{1 - \delta_l}}, \quad d = \frac{c}{1 - \delta_r + \delta_r \cdot c}.$$

It should be noted that $0 < c, d < 1$, given $0 < \delta_l, \delta_r < 1$.

5.2. Implementation of Spectrum Sharing Bargaining Scheme with Incomplete Information

The inter-operator spectrum sharing scheme also runs periodically. In each period, the procedure includes the following three stages shown in Figure 3.

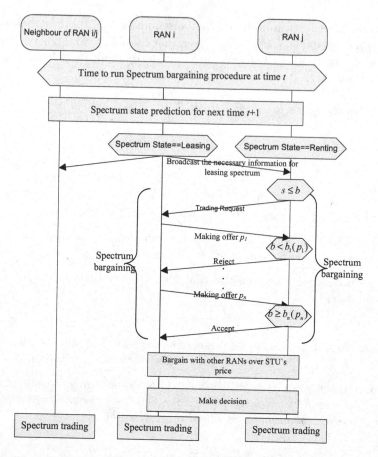

Figure 3: The message sequence chart of Inter-operator Spectrum Sharing Bargaining Scheme.

Stage 1: Pre-Bargaining

When it is time to run spectrum bargaining procedure at time t, each RAN predicts his spectrum utilisation ratio, r_s, according to Eq. (4.6). Similar to that in intra-operator spectrum sharing scheme, all RANs also are classified into two categories, i.e., the leasing RAN and the renting RAN with respect to the spectrum utilisation ratio.

And then, all TAs calculate the valuation of each STU (say b for TA_r and s for TA_l) and the discount factor according to Eq. (4.4).

Finally, all the TA_ls broadcast the necessary information, which includes the discount factor, δ_l, and the STU's valuation, s, on the CPC to inform that they want to lease out their adequate STUs at time $t + 1$.

Stage 2: Bargaining

Each TA_r selects a TA_l to bargaining over the STU's price if $b \geq s$. The TA_r sends a request message to the selected TA_l, which includes the discount factor, δ_r, the valuation distribution density, $f(b)$, and the valuation range, $[\underline{b}, \overline{b}]$.

The TA_l makes all offers, p_n, according to the Eq. (5.1), and TA_r accepts the offer if and only if $b \geq b_n(p_n)$.

According to the above theoretic analysis, there must exist a price p_n with which both TA_l and TA_r satisfy.

Stage 3: Post-bargaining

All TA_l weigh all the potential bargaining contracts and select the most profitable TA_r to trade the spectrum.

6. Simulations and Analysis

6.1. Simulation Scenario

In the simulation, three overlaid RANs, DVB-T, UMTS, and GSM, are employed to model the reconfigurable systems. Under this environment, we compare the performance of the proposed two bargaining based spectrum sharing schemes with that of the FSM method over a span of 24 hours. As adopted in [2], the spectrum sharing schemes are executed every half an hour. In order to avoid the influence of physical technologies such as channel coding and modulation scheme, we measure the spectrum utilisation instead of spectrum efficiency. Moreover, the network revenue and network call blocking rate are also measured to evaluate the performance of the three spectrum management schemes.

For illustrative purpose, various service demands for spectrum (e.g., voice load, video load, data load and etc.) across different RANs are abstracted in STUs. We assume that in the VSM, the whole block of spectrum is in the size of 190 STUs, where UMTS, GSM and DVB-T own 50, 40 and 100 STUs, respectively.

In the simulation, the advance payment parameter is set to 0.1 and the threshold Th is set to 1, i.e., no resource reservation. As the inter-operator spectrum sharing scheme need evaluate the STU's valuation, we have the following rules for RANs to value the STUs. For leasing RAN, since the surplus STU will not generate any revenue, the valuations of these STUs are set to 0. Instead, the STUs in renting RAN are used to provide service, thus the valuation of these STUs is set to the corresponding RAN's service price. Furthermore, for the sake of simplicity, the valuation distribution of renting RAN's STU is assumed to be uniform on $[\underline{b}, \overline{b}]$, so that Eqs. (5.2) can be used to calculate price offer, p_n, for TA_l and indifference valuation, b_n, for TA_r at each negotiation round. Table 1 lists the main simulation parameters as well as other network-dependent parameters.

Referring to [3, 17, 18], double-Gaussian and trapezoidal functions are adopted to simulate the historical statistical load pattern of UMTS and GSM, respectively. And the curve of DVB-T history derives from Kiefl [18].

The traffic load distributions of the three RANs over a span of 24 hours are shown in Figure 4. Since UMTS, GSM and DVB-T have different time-varying

Table 1: Simulation Parameters

RAN	DVB-T	UMTS	GSM
K	100	50	40
λ	1	3	2
s	0	0	0
b	1	3	2
$[\underline{b}, \overline{b}]$	$[0.5, 1.5]$	$[2.5, 3.5]$	$[1.5, 2.5]$
γ		0.1	
Th		0	

loads, the spectrum sharing among the three RANs are possible on a temporal basis.

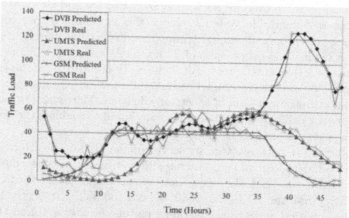

Figure 4: Three networks' traffic load (predicted and real).

As aforementioned, the spectrum sharing schemes running at t need predict the traffic load for time $t + 1$. The curves marked as "predicted" in Figure 4 are the traffic load used by spectrum sharing schemes. In view of unexpected load peaks or slopes caused by certain events, e.g., traffic jam, important sports events or a public holiday, the real traffic load values, marked as "real", will have some deviations from the predicted values in Figure 4. The performance of various spectrum sharing schemes are measured under real traffic load values.

6.2. Simulation Results

At first, we investigate the revenue of DVB-T, UMTS and GSM with inter-operator spectrum sharing scheme, intra-operator spectrum sharing scheme and FSM in Figure 5 to Figure 7. It is shown that all the three RANs definitely make more

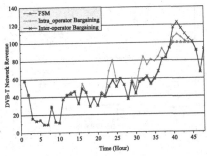

Figure 5: DVB-T network revenue.   Figure 6: UMTS network revenue.

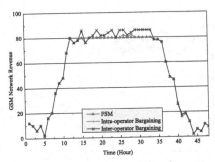

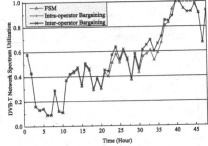

Figure 7: GSM network revenue.   Figure 8: DVB-T network spectrum utilisation.

profits with our proposed two spectrum sharing schemes, compared with conventional FSM. The differences lie in that spectrum sharing schemes allow for the spectrum trading and negotiation, which are not available in FSM. Consequently, in spectrum sharing schemes whenever any RANs have difficulty in supporting their prospective service demands, they will bargain to trade spectrum with the others. Especially for DVB-T, during UMTS and GSM busy hours, the increase of his profits is approximately up to 20% on average.

Figure 8 to Figure 13 show the spectrum utilisation and call blocking rate of the three networks under all spectrum sharing schemes, respectively. From simulation time 20 to 35, some additional spectrums are needed by UMTS and GSM network, while the spectrums are adequate in DVB-T network. As a result of the proposed spectrum sharing schemes, the DVB-T leases out its adequate STUs to the other two networks, which consequently increase the spectrum utilisation of DVB-T (see Figure 8) and decrease blocking rate of UMTS and GSM network (see Figure 12 and Figure 13). However, since the spectrum requirements in UMTS and GSM network are small (less than 10 STUs), the improvement of spectrum utilisation of DVB-T is 10% at most. Meanwhile, the spectrum utilisation of UMTS

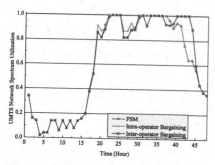

Figure 9: UMTS network spectrum utilisation.

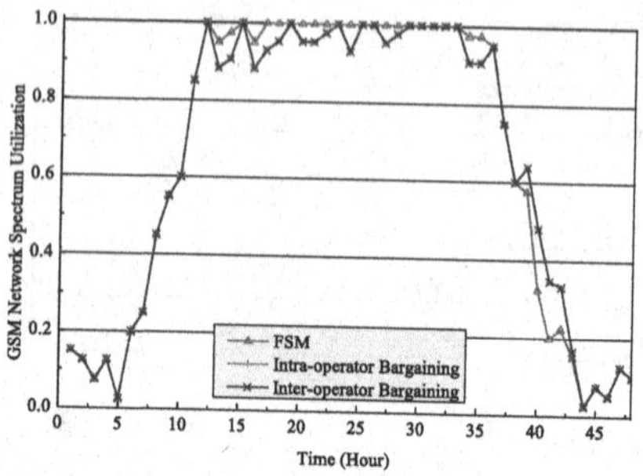

Figure 10: GSM network spectrum utilisation.

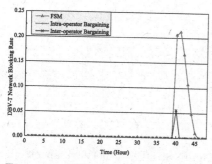
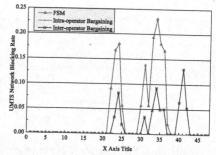

Figure 11: DVB-T network blocking rate.

Figure 12: UMTS network blocking rate.

and GSM network is still maintained above 90%. Note that the small jitters of GSM network's spectrum utilisation (see Figure 13) are caused by the deviation between predicted traffic load and real traffic load.

From simulation time 40 to 45, see Figure 4, the DVB-T network becomes lacking in STUs, and rents STUs from UMTS and GSM networks with the constraints of inter-operator and intra-operator spectrum sharing schemes. Consequently, both UMTS and GSM network's spectrum utilisation increase (see Figure 9 and Figure 10). Especially, the improvement of UMTS network spectrum utilisation is nearly 20% by employing our spectrum sharing schemes. However, due to the inaccurate traffic load prediction, the DVB-T network's spectrum utilisation decreases a little (around 5%). But the call blocking rate of DVB-T network decreases greatly from 20% to 5% (see Figure 11).

Finally, with respect to inter-operator spectrum sharing scheme, since the two involving RANs negotiate over the price of traded STU, we investigate the efficiency of this negotiation procedure. The cumulative probability distribution of bargaining rounds is presented in Figure 14. We notice that 90% bargaining

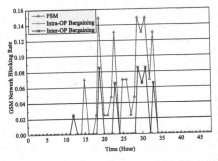

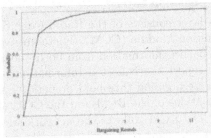

Figure 13: GSM network blocking rate.

Figure 14: The probability distribution of bargaining rounds in inter-operator spectrum sharing scheme.

procedures finished in 3 rounds. As it is a large scale (every half an hour) spectrum management method, the inter-operator spectrum sharing scheme taking 3 rounds negotiation to reach an agreement can be accepted.

7. Implementation of Trading Agent

This part discusses the internal architecture of a TA in detail. Figure 15 shows the functional architecture of a TA, which is divided into three layers, i.e., the Cognition Layer, the Reaction Layer and the Memory Layer.

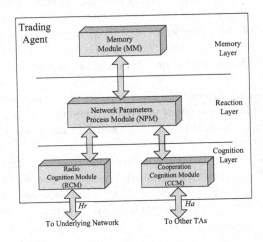

Figure 15: The functional architecture of the TA.

7.1. Cognition Layer

This layer consists of the Radio Cognition Module (RCM) and the Cooperation Cognition Module (CCM). There are two main functionalities in this layer. One is to collect information from both underlying network and other TAs. The other is to report the collected information to upper layer once the statistic exceeds its corresponding threshold or periodic timer expires. The detailed functionality descriptions of the RCM and the CCM are given as follows.

The RCM collects the network status information, such as the traffic load, the call blocking probability, from the underlying network via the Hr interface and reports it to the upper layer based on a certain criteria. The criteria are also set by the upper layer, and may be threshold-based or timer-based. In addition, the RCM relays those network operational parameters generated by the upper layer to the underlying network via Hr interface.

Similarly, the CCM executes the similar operations like the RCM including collecting status information and reporting to the upper layer. The only difference lies in that CCM collects the status information from other TAs via the Ha interface instead of the Hr interface. Particularly, CCM must have the important capability of discovering the neighbouring TAs, which is a prerequisite for information collection.

7.2. Reaction Layer

The reaction layer is the core component of an agent, which consists of a Network parameters Process Module (NPM). The NPM implements the main tasks of the reaction layer, which include tuning the underlying network's operational parameters, i.e., operational spectrum range in this paper, setting report criteria, and interacting with the upper layer.

The first task of the reaction layer is to manage the spectrum resource autonomically as depicted above. After receiving the information from the cognition layer, the NPM may decide to rent or lease the spectrum, to achieve the predefined objectives.

The second task of the reaction layer is to set the report criteria for cognition layer. The criteria may be a set of statistics thresholds or a period of a timer. Thus, the cognition layer delivers the collected information to the reaction layer according to the criteria.

The third task is to exchange the network information with the memory layer. For each successful adjustment of network spectrum, the NPM should store the information to the memory layer. When the similar case appears, the NPM can directly read the solutions from the memory layer instead of recalculation.

7.3. Memory Layer

The Memory Module (MM) is adopted to accelerate the adjustment and achieve intelligent control. For each successful adjustment, the MM stores the corresponding network parameters. When the NPM in reaction layer faces similar situation, the most suitable network operational parameters can be obtained directly from MM rather than recalculation.

8. Conclusion

In order to facilitate the autonomic network resource management, we investigate novel bargaining based dynamic spectrum sharing schemes among RANs belonging to the same operator and multi-operators. By providing the multi-agent architecture for spectrum sharing, we propose the spectrum blocks called STUs be traded among RANs according to their negotiated results. With the help of bargaining game theory, the negotiation procedure between intra-operator RANs are simplified to reach an agreement immediately. And the negotiation procedure between inter-operator RANs is proved to be convergent and the convergence path is also given. Simulation results demonstrate that our approaches remarkably outperform the existing FSM in expanding the network's revenue, decreasing call blocking rate and bettering the efficiency of spectrum.

Since the information acquirement between operators is a main challenge for spectrum bargaining, we propose spectrum bargaining scheme under one sided uncertainty. For future works, we will further investigate the bargaining procedure under two sided uncertainty.

References

[1] P. Leaves, K. Moessner, R. Tafazolli, *Dynamic spectrum allocation in composite reconfigurable wireless networks.* IEEE Communications Magazine, 2004, 72–81.

[2] P. Leaves et al., *Dynamic spectrum allocation in hybrid networks with imperfect load prediction.* 3G Mobile Communication Technologies, May 2002, 444–448.

[3] P. Leaves et al., *Dynamic spectrum allocation in a multi-radio environment: concept and algorithm.* 3G Mobile Communication Technologies, March 2001, 53–57.

[4] White Paper Draft, *Cognitive radio, spectrum and radio resource management.* SDR community within WWRF.

[5] A. Rubinstein, M. Osborne, *Bargaining and Markets.* Academic Press 1990.

[6] D. M. Kreps, R. Wilson, *Sequential Equilibria.* Econometrica, **50**, 1982, 863–894.

[7] D. Fudenberg, D. K. Levine, J. Tirole, *Infinite-Horizon Models of Bargaining with One-Sided Incomplete Information.* Levine's Working Paper Archive 1098, UCLA Department of Economics, 1985.

[8] X. Jing, D. Raychaudhuri, *Spectrum co-existence of IEEE 802.11b and 802.16a networks using CSCC etiquette protocol.* IEEE DySPAN, Nov. 2005, 243–250.

[9] O. Ileri, D. Samardzija, N. B. Mandayam, *Demand responsive pricing and competitive spectrum allocation via spectrum server.* IEEE DySPAN, Nov. 2005, 194–202.

[10] G. Marias, *Spectrum scheduling and brokering based on QoS demands of competing WISPs.* IEEE DySPAN, Nov. 2005, 684–687.

[11] L. Cao, H. Zheng, *Distributed spectrum allocation via local bargaining.* IEEE Sensor and Ad Hoc Communications and Networks (SECON), Sept. 2005, 475–486.

[12] J. Zhao, H. Zheng, G.-H. Yang, *Distributed coordination in dynamic spectrum allocation networks.* IEEE DySPAN, Nov. 2005, 259–268.

[13] V. Brik, E. Rozner, S. Banarjee, P. Bahl, *DSAP: a protocol for coordinated spectrum access.* IEEE DySPAN, Nov. 2005, 611–614.

[14] N. Nie, C. Comaniciu, *Adaptive channel allocation spectrum etiquette for cognitive radio networks.* IEEE DySPAN, Nov. 2005, 269–278.

[15] K. Moessner, *E2R Regulatory Perspectives.* WWI Symposium, Dec. 2004.

[16] P. Cordier et al., *E2R Cognitive Pilot Channel concept.* IST, 2006.

[17] S. Almeida, J. Queijo, L. Correia, *Spatial and temporal traffic distribution models for GSM.* Vehicular Technology Conference, **1**, Sept. 1999, 131–135.

[18] B. Kiefl, *What will we watch? A Forecast of TV viewing habits in 10 years.* The Advertising Research Foundation, New York, USA, 1998.

Acknowledgement

This work has been performed in the framework of the EU funded project E^2R II (IST-2005-027714). It's also supported by National Natural Science Foundation of China (60632030), China-EU S&T Cooperation Foundation of Ministry of S&T of China (0516) and National 863 High Tech R&D Program of China, (2006AA01Z276). The authors would like to acknowledge the contributions of their colleagues from E^2R II consortium and from Wireless Technology Innovation Institute of BUPT.

Jie Chen
P.O. Box 92
Beijing University of Posts and Telecommunications
10 Xi Tu Cheng Road, Haidian District, Beijing, China
e-mail: jie.chen.bupt@gmail.com

Miao Pan, Kai Yu, Yang Ji and Ping Zhang
P.O. Box 92
Beijing University of Posts and Telecommunications
10 Xi Tu Cheng Road, Haidian District, Beijing, China
e-mail: {lamb0327, kai.yu.bupt}@gmail.com
        {jiyang, pzhang}@bupt.edu.cn

Whitestein Series in Software Agent Technologies and Autonomic Computing

Edited by
Marius Walliser, Stefan Brantschen, Monique Calisti and Stefan Schinkinger

This series reports new developments in agent-based software technologies and agent-oriented software engineering methodologies, with particular emphasis on applications in the area of autonomic computing and communications.
The spectrum of the series includes research monographs, high quality notes resulting from research and industrial projects, outstanding Ph.D. theses, and the proceedings of carefully selected conferences. The series is targeted at promoting advanced research and facilitating know-how transfer to industrial use.

■ **Calisti, M.**, Whitestein Technologies AG, Zürich, Switzerland / **van der Meer, S.**, Waterford Institute of Technology, Ireland / **Strassner, J.**, Motorola, Inc., Schaumburg, IL, USA (eds.)

Advanced Autonomic Networking and Communication

This book presents a comprehensive reference of state-of-the-art efforts and early results in the area of autonomic networking and communication.
The essence of autonomic networking, and thus autonomic communications, is to enable the self-governing of services and resources within the constraints of business rules. In order to support self-governance, appropriate self-* functionality will be deployed in the network on an application-specific basis. The continuing increase in complexity of upcoming networking convergence scenarios mandates a new approach to network management.
This volume explores different ways that autonomic principles can be applied to existing and future networks. In particular, the book has 3 main parts, each of them represented by three papers discussing them from industrial and academic perspectives.
The first part focuses on architectures and modeling strategies. Part two is dedicated to middleware and service infrastructure as facilitators of autonomic communications, and the last part addresses autonomic networks, specifically how current networks can be equipped with autonomic functionality and thus migrate to autonomic networks.

2007. 200 pages. Softcover.
ISBN 978-3-7643-8568-2

■ **Annicchiarico, R.**, Fondazione Santa Lucia IRCCS, Rome, Italy / **Cortés, U.**, Universidad Malaga, Spain / **Urdiales, C.**, Universidad Polytècnica de Catalunya, Barcelona, Spain (eds.)

Agent Technology and e-Health

2007. 156 pages. Softcover.
ISBN 978-3-7643-8546-0

■ **Pěchouček, M.**, Czech Technical University, Prague, Czech Republic / **Thompson, S.G.**, BT. Labs, Suffolk, U.K. / **Voos, H.**, University of Applied Sciences, Ravensburg-Weingarten, Germany (eds.)

Defense Industry Applications of Autonomous Agents and Multi-Agent Systems

Defense and security related applications are increasingly being tackled by researchers and practioners using technologies developed in the field of Intelligent Agent research. This book is a collection of recent refereed papers drawn from workshops and other colloquia held in various venues around the world in the last two years.
The contributions in this book describe work in the development of command and control systems, military communications systems, information systems, surveillance systems, autonomous vehicles, simulators and Human Computer Interactions. The broad nature of the application domain is matched by the diversity of techniques used

in the papers that are included in the collection which provides, for the first time, an overview of the most significant work being performed by the leading workers in this area. It provides a single reference point for the state of the art in the field at the moment and will be of interest to Computer Science professionals working in the defense sector, and academics and students investigating the technology of Intelligent Agents that are curious to see how the technology is applied in practice.

2007. 180 pages. Softcover.
ISBN 978-3-7643-8570-5

■ **Moreno, A.** University of Tarragona, Spain / **Pavón, J.**, University of Madrid, Spain (eds.)

Issues in Multi-Agent Systems
The AgentCities.ES Experience

The agent paradigm has been a subject of research for the last years, and the purpose of this book is to present current status of this technology by looking at its application in different domains, such as electronic markets, e-tourism, ambience intelligence, and complex system analysis.

2007. 240 pages. Softcover.
ISBN 978-3-7643-8542-2

■ **Pautasso, C.**, IBM Zürich, Switzerland / **Bussler, C.**, Cisco Systems Inc., San Jose, USA (eds.)

Emerging Web Services Technology

2007. 182 pages. Softcover.
ISBN 978-3-7643-8447-0

Whitestein Series in Software Agent Technologies and Autonomic Computing

Edited by

Marius Walliser, Stefan Brantschen, Monique Calisti and Stefan Schinkinger

This series reports new developments in agent-based software technologies and agent-oriented software engineering methodologies, with particular emphasis on applications in the area of autonomic computing and communications.
The spectrum of the series includes research monographs, high quality notes resulting from research and industrial projects, outstanding Ph.D. theses, and the proceedings of carefully selected conferences. The series is targeted at promoting advanced research and facilitating know-how transfer to industrial use.

■ **Cervenka, R. / Trencansky, I.,** both Whitestein Technologies, Bratislava, Slovakia

The Agent Modeling Language - AML. A Comprehensive Approach to Modeling Multi-Agent Systems

Modeling of multi-agent systems still lacks complete and proper definition, general acceptance, and practical application. Due to the vast potential of these systems e.g., to improve the practice in software and to extent the applications that can feasibly be tackled, this book tries to provide a comprehensive modeling language (AML) as an extension of UML 2.0, concentrating on multi-agent systems and applications.

2007. 366 pages. Softcover.
ISBN 978-3-7643-8395-4

■ **van Dinther, C.,** Karlsruhe, Germany

Adaptive Bidding in Single-Sided Auctions Under Uncertainty. An Agent-based Approach in Market Engineering

2006. 256 pages. Softcover.
ISBN 978-3-7643-8094-6

This book shows that and how software agents can be used to simulate bidding behaviour in electronic auctions. The main emphasis of this book is to apply computational economics to market theory. It summarizes the most common and up-to-date agent-based simulation methods and tools and develops the simulation software AMASE. On basis of the introduced methods a model is established to simulate

bidding behaviour under uncertainty.

■ **Zimmermann, R.,** Erlangen, Germany

Agent-based Supply Network Event Management

2006. 340 pages. Softcover.
ISBN 978-3-7643-7486-0

■ **Unland, R.,** Essen, Germany / **Klusch, M.,** Saarbrücken, Germany / **Calisti, M.,** Zürich, Switzerland (eds.)

Software Agent-based Applications, Platforms and Development Kits

2005. 462 pages. Softcover.
ISBN 978-3-7643-7347-4

■ **Klügl, F.,** Würzburg, Germany / **Bazzan, A.,** Porto Alegre, Brazil / **Ossowski, S.,** Madrid, Spain (eds.)

Applications of Agent Technology in Traffic and Transportation

2005. 218 pages. Softcover.
ISBN 978-3-7643-7258-3

■ **Tamma, V.,** Liverpool, U.K. / **Cranefield, S.,** Dunedin, New Zealand / **Finin, T.W.,** Baltimore, U.S.A. / **Willmott, S.,** Barcelona, Spain (eds.)

Ontologies for Agents: Theory and Experiences

2005. 356 pages. Softcover.
ISBN 978-3-7643-7237-8

■ **Neagu, N.,** Zürich, Switzerland

Constraint Satisfaction Techniques for Agent-Based Reasoning

2005. 172 pages. Softcover.
ISBN 978-3-7643-7217-0

■ **van Aart, C.,** Waalwijk, The Netherlands

Organizational Principles for Multi-Agent Architectures

2005. 216 pages. Softcover.
ISBN 978-3-7643-7213-2

■ **Vázquez-Salceda, J.,** Utrecht University, The Netherlands

The Role of Norms and Electronic Institutions in Multi-Agent Systems

2004. 292 pages. Softcover.
ISBN 978-3-7643-7057-2

■ **Moreno, A.,** Tarragona, Spain / **Nealon, J.L.,** Oxford, U.K. (eds.)

Applications of Software Agent Technology in the Health Care Domain

2003. 212 pages. Softcover.
ISBN 978-3-7643-2662-3

■ **Calisti, M.,** Zürich, Switzerland

An Agent-Based Approach for Coordinated Multi-Provider Service Provisioning

2002. 292 pages. Softcover.
ISBN 978-3-7643-6922-4

■ **Günter, M.,** Zürich, Switzerland

Customer-based IP Service Monitoring with Mobile Software Agents

2002. 168 pages. Softcover.
ISBN 978-3-7643-6917-0

BIRKHÄUSER